FUNDAMENTALS OF SCIENTIFIC MATHEMATICS

Fundamentals of Scientific Mathematics

By George E. Owen
Professor of Physics, The Johns Hopkins University

THE JOHNS HOPKINS PRESS   BALTIMORE, MARYLAND

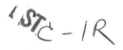

THIS BOOK WAS PUBLISHED WITH THE ASSISTANCE OF
A GRANT FROM THE ESSO EDUCATION FOUNDATION.

ILLUSTRATED BY THE AUTHOR

Distributed in Great Britain by Oxford University Press, London
Printed in the United States of America
Library of Congress Catalog Card Number 61-12287

DEDICATED TO MY WIFE KARICIĞIM DEHA

Preface

IN THE SUMMER of 1958 The Johns Hopkins University was presented with a unique opportunity to experiment in the field of mathematics teaching at the secondary-school level. This opportunity was given to us by the Esso Education Foundation. Aside from expressing an interest in the improvement of science and mathematics teaching, the Foundation granted the University a free hand to do what it considered to be the more significant investigation.

After a series of consultations involving various members of the science faculties and the administration, a decision was reached to investigate the field intermediate between pure mathematics and the natural sciences. This field was that of applied mathematics or scientific mathematics.

The actual program took the form of a special in-service training program for a selected group of secondary-school teachers. During the two academic years 1958–1959 and 1959–1960, one-hundred secondary-school mathematics and science teachers from the city of Baltimore and Baltimore County participated in the course. These participants have had a marked influence on this book as it now appears in final form.

Together with Dr. Joseph Sampson of the Department of Mathematics, the author took the major responsibility for the course to be presented. We were aided from the beginning by the advice and support of two committees—one from The Johns Hopkins University, and one composed of the science and mathematics supervisors from Baltimore and Baltimore County public and private schools.

The author wishes to acknowledge the contributions and support of many individuals whose efforts contributed markedly to this book.

I wish to express my appreciation for the generous support of the Esso Education Foundation and in particular of the enthusiastic interest displayed by Mr. Claude Alexander, former Secretary of the Foundation; Dr. C. L. Brown, Manager, Scientific Liaison of the Esso Research and Engineering Company; Mr. William P. Headden of the Public Relations Department, Standard Oil Company of New Jersey; and Mr. George M. Buckingham, present Secretary of the Esso Education Foundation.

I am indebted to Dr. Joseph Sampson who contributed both to the course notes and to the lectures. Dr. Sampson's help and advice have formed a sizeable portion of the final result.

The initial idea for the teaching program came from Mr. P. Stewart Macaulay, Executive Vice-President of The Johns Hopkins University. When the question was first raised concerning the type of study which might be given, Mr. Macaulay proposed that a course be established to develop instruction in scientific mathematics. I am indeed grateful for his contributions.

The author particularly thanks Mr. Wilbert E. Locklin, Assistant to the President of The Johns Hopkins University, who provided constant assistance and encouragement throughout the project.

I thank Mrs. Dencie Kent for her careful typing of the manuscript, Mr. P. W. Keaton who assisted in the course, and both the University committee and the committee from the secondary schools.

Baltimore, January, 1961

GEORGE E. OWEN

Contents

CHAPTER 4 FUNCTIONS

CHAPTER 5 DIFFERENTIAL AND INTEGRAL CALCULUS

FUNDAMENTALS OF SCIENTIFIC MATHEMATICS

Notation

Vectors which are designated by a symbol such as $\vec{A}$ in the illustrations will be represented by bold face letters such as **A** in the text.

Matrices will be represented by open face letters such as $\mathbb{S}$ in the text. The elements of a matrix are shown as symbols having two subscripts.

Punctuation at the end of equations has been omitted throughout in the interest of clarity.

CHAPTER 1

Geometry and Matrices

A. Description of E_3

EVERYONE IS ENDOWED to some extent, at least, with an understanding of the physical ***space*** that surrounds us and the events of nature that take place in that space. Such events as the motion of material bodies, propagation of sound waves, and chemical and biological changes are in many cases perceptible. It is the business of the scientist to describe events of this type as accurately as he can.

A description of natural events can be phrased in ordinary language, of course, and ***before*** Newton's time nearly all scientific observations were cast in that form. But verbal language has been found to be extremely unwieldy and unsuited to accurate descriptions of natural phenomena.

The scientist, therefore, must use the language of Mathematics. In order to use this flexible and more precise language ***he must represent the phenomena of nature by mathematical objects,*** which correspond to observations in some manner. In the pages to follow we shall give an account of how this is done as well as an account of some very important mathematical concepts.

A given set of mathematical objects is sometimes like the parts of a game. Symbols, pictorial or abstract, are supplied the contestant. He is then provided with a strict set of rules for manipulating the symbols. The important aspect of the game is that once the rules have been set they can never be violated at any extension of the game.

Very often it is difficult to decide just how to represent a situation in nature by a mathematical object. The choice of the mathematical object is never unique, and the usefulness of a given choice depends largely upon the *skill* and *insight* of the scientist. *However, a large body of experience has established the usefulness and accuracy of certain standard mathematical objects as sufficient representatives for a variety of real things.*

Among these useful objects is the ***three-dimensional Euclidean space*** (we shall call it simply E_3). E_3 is utilized as a mathematical representative of a **real local space.**

We use the word **local** because even at first thought we should not expect to extrapolate the properties of the space observed in our immediate vicinity to the entire space of the universe. More will be said about the concept of a local space later in this book.

Asserting that a two-dimensional space is represented by the plane of this page (called E_2), then E_3 is simply a three-dimensional version of E_2. In other words, besides including all points in the plane of this page we include all points in parallel planes above and below.

The student will recall how E_2 is described by means of certain

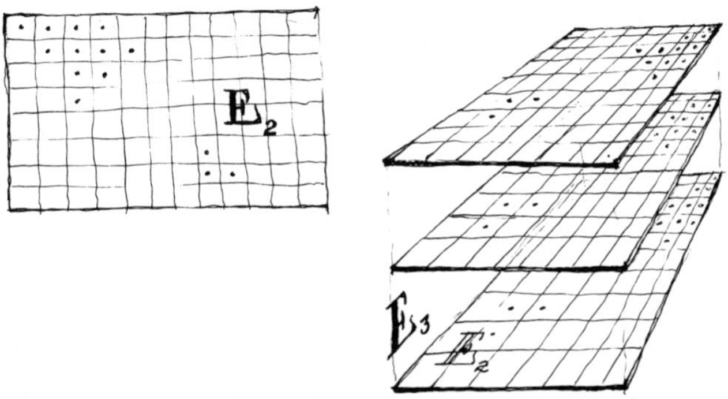

axioms. E_3 is also a system described by axioms, which we shall give presently.

The axioms which we shall use are in the form best suited to our purposes and are somewhat different in appearance from the familiar axioms of plane geometry.

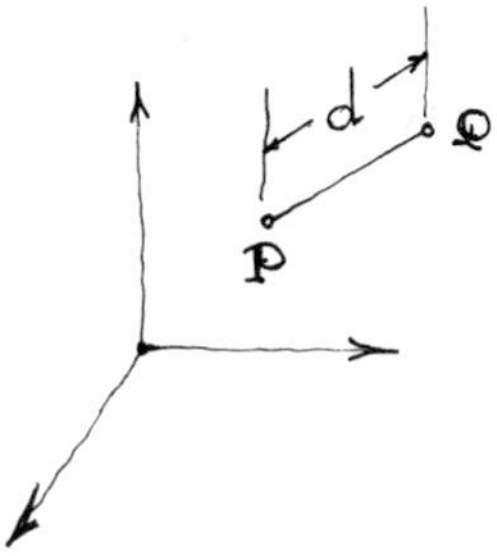

E_3 is first of all a collection of (undefined) objects called points. These points are imagined as corresponding to the points of the ***real local space.*** The axioms of E_3 assign to every pair of points P, Q a **number** called the **distance** between P and Q. This ***distance*** is supposed to represent the physical distance between the real points corresponding to P and Q (measured in appropriate units).

Certain collections of points in E_3 will be called **straight lines.** These will be defined later. They are used to represent the straight lines of the physical space.

We point out here that we can define the straight lines in a real local space as the paths of light rays. This is to say the straight line joining two real points is by definition the line of sight from one point to the other.

It is of some interest to note that the straight lines of E_3 for most problems will correspond to the light rays of the physical space to a high degree of accuracy. For instance if we represent the refraction of light in an optical lens by lines in E_3 we obtain a very ac-

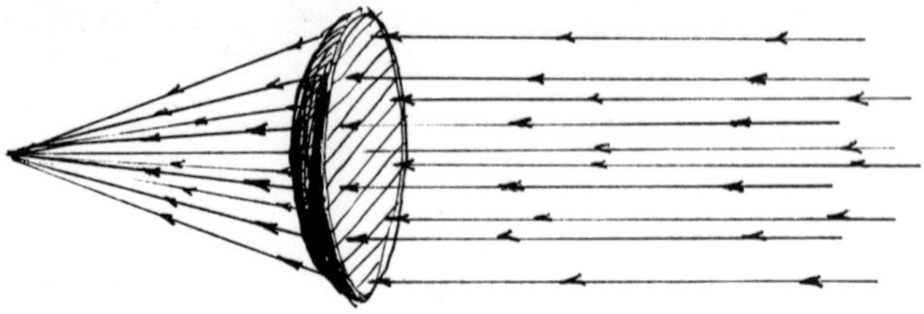

curate correspondence between the diagram in E_3 and the physical results in our local space.

On the other hand there is reason to believe that the paths of light rays will be affected by the presence of material bodies. The propositions of General Relativity suggest that for astronomical distances the straight lines of E_3 will not correspond to light rays.

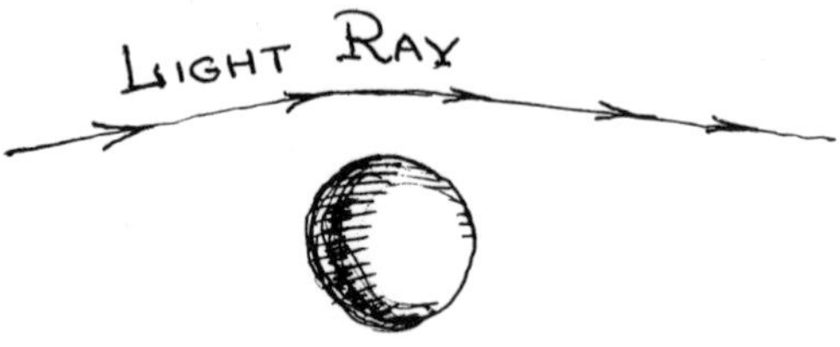

In spite of the limitations of E_3 when applied to problems which deal with astronomical distances, we can use E_3 as a mathematical model of our local space to a high degree of accuracy.

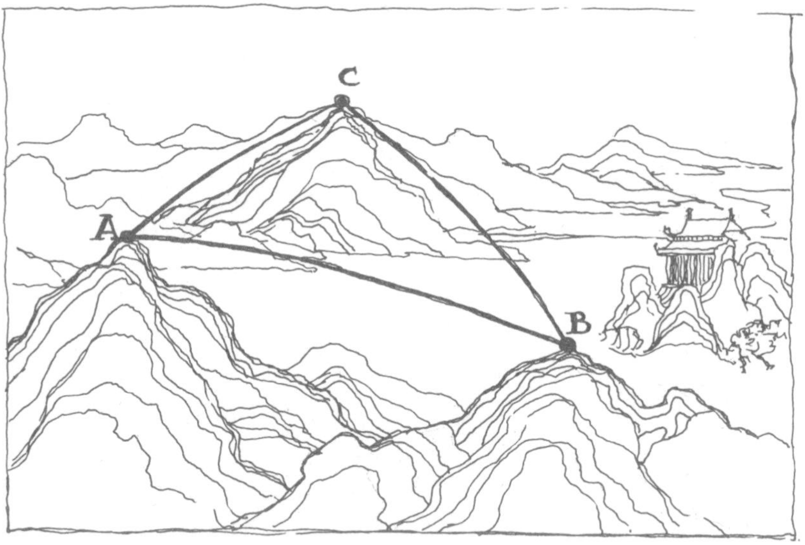

In the nineteenth century, C. F. Gauss proposed an experiment to test the validity of E_3.

Consider a physical triangle whose vertices A, B, and C lie on three mountain peaks. The sides of the triangle are then the lines of sight joining the three points in question. Suppose now that we measure the three angles of our triangle as accurately as possible.

There is reason to suppose that the sum of the three angles will not add exactly to 180°. The departure of the sum from 180° would provide us with a measure of the validity of E_3 for such distances as are involved.

Interestingly enough, such measurements add to 180° within the experimental error of present day measurements. Thus E_3 is adequate for most local problems.

Once a departure from E_3 is detected, a new or modified mathematical model must be substituted whenever great accuracy is required. Such a model must behave as E_3 in those approximations in which E_3 is known to provide a description.

B. The Axioms of E_3

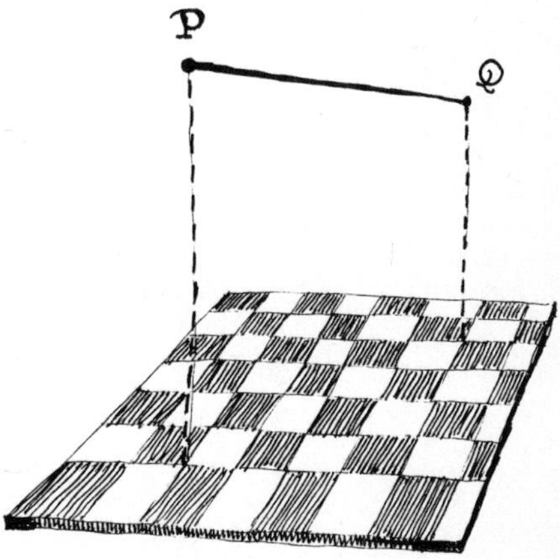

A precise description of E_3 can be presented by stating the axioms of E_3. As has been stated previously, E_3 is composed of a set of abstract objects called ***points.***
The axioms of E_3 provide a relationship between these points.

Many equivalent axioms can be given. For our purposes the following two axioms are the simplest:

Axioms of E_3

1. For any two points P, Q there is assigned a number d, where $d \geq 0$. This number is called the distance between P and Q.

2. It is possible to assign to every point of E_3 three numerical coordinates x, y, and z such that:

(a) For any triplet of numbers (x, y, z) there is one and only one point of E_3 having (x, y, z) as its coordinates.

(b) If P has the coordinates (x, y, z) and if Q has the coordinates (x′, y′, z′) then d, the distance between P and Q, is equal to

$$[(x - x')^2 + (y - y')^2 + (z - z')^2]^{1/2}$$

These two axioms contain all that we shall need to know about E_3. Since the points in E_3 are thought of as corresponding to the points of physical space in such a way that if P and Q in E_3 correspond to P_0 and Q_0 in real space, then d, the distance between P and Q, is equal numerically to the distance between P_0 and Q_0.

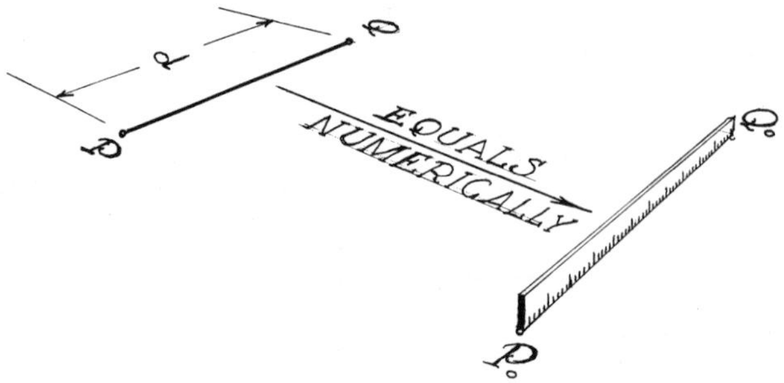

Before discussing the coordinate system we should note that the definition of the distance between two points (x, y, z) and (x′, y′, z′) implies the Pythagorean theorem. To be more specific our definition of d requires that the line (x − x′), the line (y − y′), and the line (z − z′) all be mutually perpendicular. The proof of this can be seen readily from the diagram below.

In the section to follow we shall find that the rule

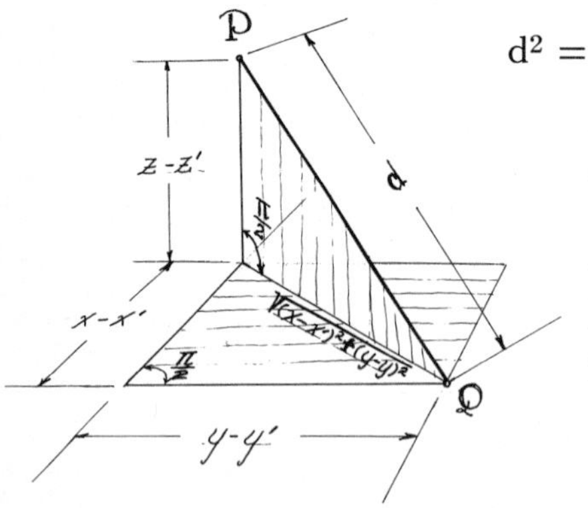

$$d^2 = (x - x')^2 + (y - y')^2 + (z - z')^2$$

will require a set of coordinates which are mutually perpendicular.

As an example of the definition of the distance between two points in E_2 let us consider the following problem.

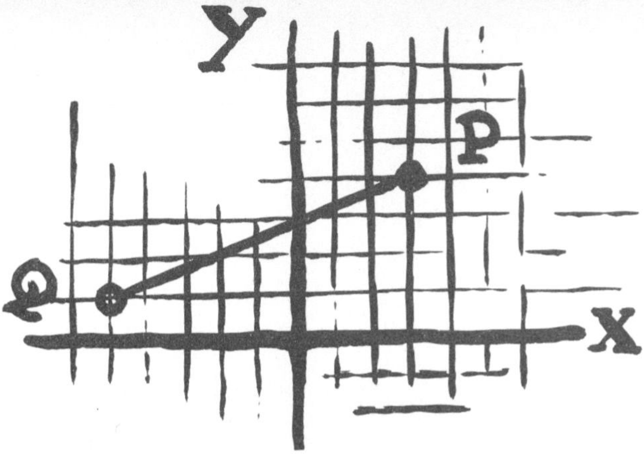

Point P is located by the coordinates (3, 4). This is to say $x = 3$ and $y = 4$.

Assume in this case point Q is given by $(-5, 1)$, i.e. $x' = -5$ and $y' = 1$.

The distance between P and Q is then

$$d = [(3 - \{-5\})^2 + (4 - 1)^2]^{1/2} = [73]^{1/2}$$

C. The Coordinate System

Orthogonal and non-orthogonal systems

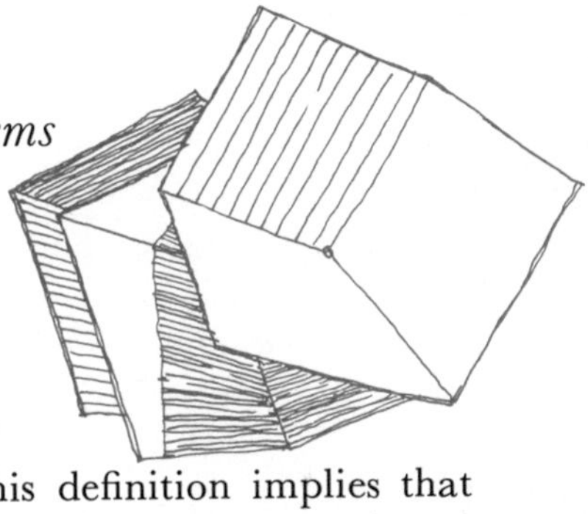

The distance "d" in E_3 have been defined without reference to a specific coordinate system.

As we have mentioned in the last section our definition of the distance "d" invokes the Pythagorean theorem. This definition implies that each of the triplet of numbers specifying a point (such as the number) x in (x, y, z) corresponds to a set of measurements along ***three mutually perpendicular axes***[1].

Utilizing our spatial intuition we can specify that the perpendicular (or orthogonal) coordinates are formed by the intersection of three mutually perpendicular planes. As an example the walls forming the corner of a room form a set of orthogonal (or perpendicular) coordinates.

[1] By perpendicular we mean that the angle between any pair of axes is 90° (measured in the plane formed by the two axes).

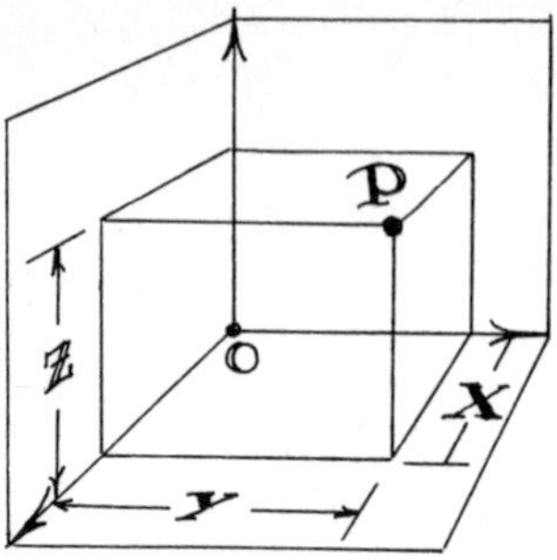

The coordinate x, is found by passing a plane through P parallel to the yz plane. The intersection between this plane and the x axis specifies the x coordinate of the point P.

In like manner a plane through P parallel to the zx plane provides the y coordinate of P and a plane through P parallel to the xy plane provides the z coordinate.

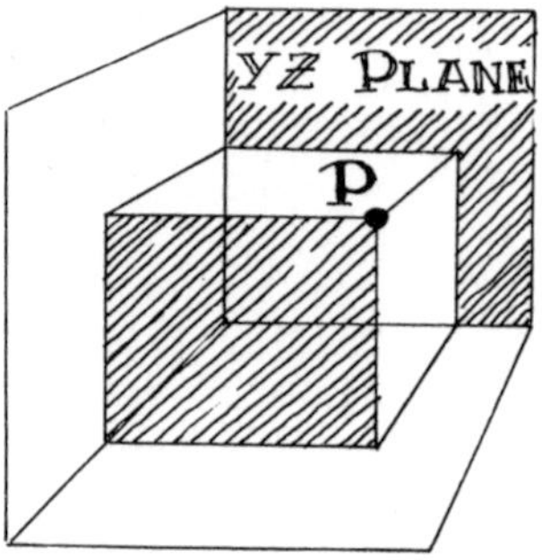

We then see that the original planes defining the coordinate axes plus the three planes defining the coordinates (x, y, z) of the point P form a rectangular parallelepiped.

The coordinate axes are defined as that collection or sequence of points whose coordinates are

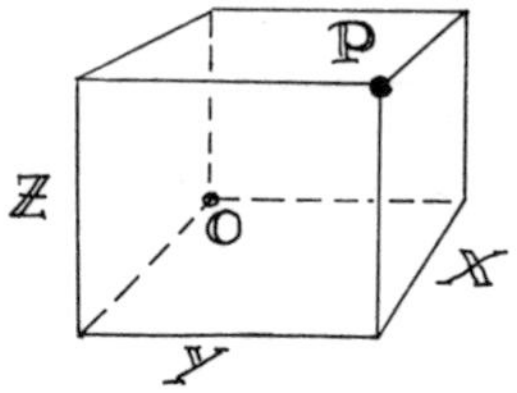

(x,o,o); **the** x **axis**
(o,y,o); **the** y **axis**
(o,o,z); **the** z **axis**

This particular system is called CARTESIAN.[2]

By passing from the points of E_3 to their coordinates we are able to translate geometrical problems into numerical problems. Geometry referred to a specific coordinate system is called ***analytic (or coordinate) geometry.***

It should be emphasized that the coordinate system is not unique; it only provides a frame for reference.

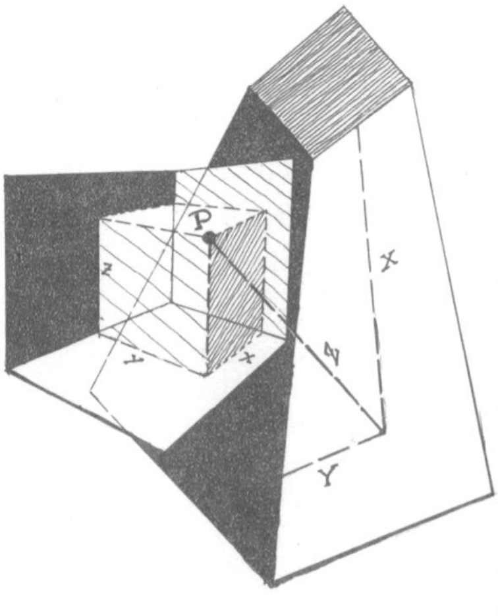

[2] In some texts the term Cartesian has a slightly broader meaning than we have given here.

P can be specified relative to the frame O or relative to a frame O′.

In the preceding diagram a second set of axes are shown designated by the origin O′. We see that the point P could have been described by the triplet of numbers (X, Y, Z).

Throughout this discussion we have emphasized the condition that our coordinate axes formed a mutually perpendicular set of axes. A set of non-orthogonal axes could have been used. It is possible to describe the point P in terms of any non-coplanar[3] set of axes. However, in doing so we should violate axiom 2 (b), which defines the distance between two points.

To illustrate this let us consider a simple two-dimensional set of axes. A coordinate system satisfying 2 (b) (the definition of d) is called orthogonal or Euclidean. A two-dimensional Euclidean system can be called E_2, where

$$d^2 = (x - x')^2 + (y - y')^2$$

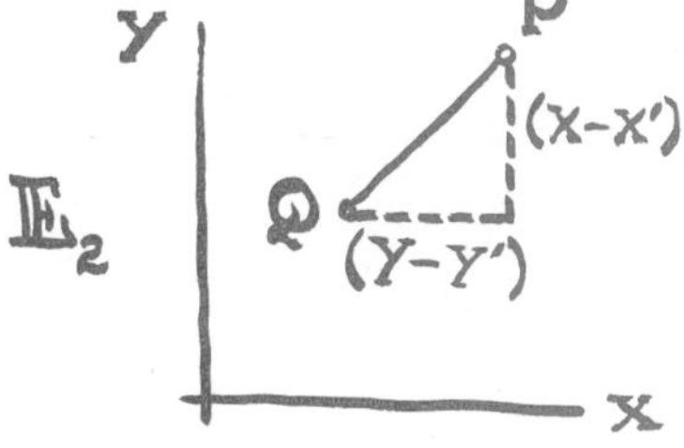

Regard the ***non-orthogonal*** two-dimensional system shown. This system is formed from two intersecting straight lines which intersect forming the angle α between them (α in this case is less than 90°).

This is an example in which the distance between P and Q is not given by the square root of the sum of the squares of the coordinate differences.

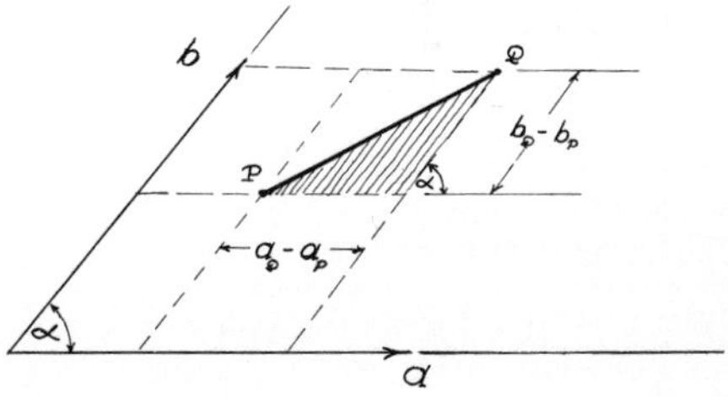

[3] By non-coplanar axes we signify that the three axes cannot lie in the same plane. This statement also provides the condition that no two axes are parallel (remembering that the axes meet or intersect at one point called the origin).

We must define the coordinates in the same manner as in the case of the orthogonal system; noting simultaneously that a plane through P in E_3 becomes a line through P in E_2.

Passing lines parallel to b through P and Q we find the intersections with the "a" axis which provide the coordinates a_p and a_q of P and Q respectively.[4]

Using lines parallel to "a" through P and Q we obtain b_p and b_q.

The distance d between P and Q is now defined in terms of the two sides of the triangle $(a_q - a_p)$, $(b_q - b_p)$, and the angle $(\pi - \alpha)$.

$$d^2 = (a_q - a_p)^2 + (b_q - b_p)^2 - 2(a_q - a_p)(b_q - b_p)\cos(\pi - \alpha)$$

It is apparent that this is a general relation since the distance "d" for E_2 is obtained in the special case in which the angle α is equal to $\pi/2$.

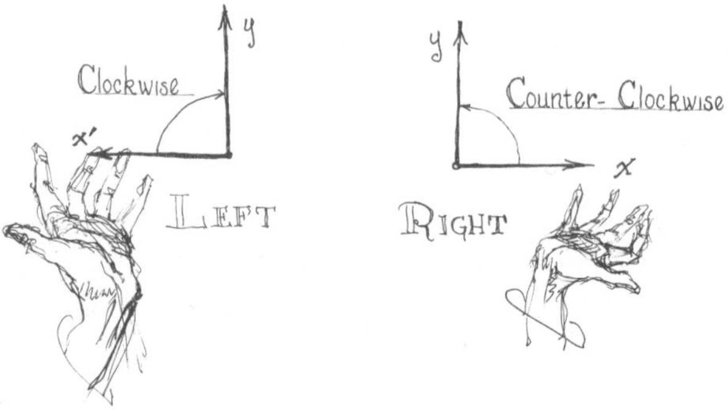

Right-handed and Left-handed Cartesian Systems

Another property of a coordinate system which must be noted is the order in which the coordinates are taken. This order determines whether a physical coordinate system is **"right handed"** or **"left handed."**

To illustrate in an elementary manner the difference between these two systems consider the standard E_2 shown at the right above. At the left is pictured a ***left handed two-dimensional space.***

In E_2 the positive x axis must be rotated **counter-clockwise** to

[4] The subscript P on "a" merely denotes that "a"$_p$ belongs to P.

obtain the positive y axis; under these conditions the system is called a ***right-handed system.***

On the other hand the ***left-handed system*** has the property that the positive x' axis must be rotated clockwise in order to obtain the positive y axis.

It is the *convention* in problems of mathematical science to choose a right-handed coordinate system.

This concept of the **"handedness"** of a set of coordinates is physiological, yet we find in physics that results of certain experiments (now labeled the parity experiments) are *not* independent of the "handedness" of our choice of laboratory coordinates.

Before discussing E_3 it should be noted that the "handedness" of our system could have been denoted by the doublet of numbers indicating a point P.

(x, y) corresponds to a right-handed system

(y, x') or $(-x, y)$ corresponds to a left-handed system

Later in the section concerning rotations the difference in the rotation properties of the two systems will give us other differences between the two systems.

In discussing the two-dimensional case the words **clockwise** and **counter-clockwise** were used. It becomes apparent then that the differentiation has a physiological aspect. We shall define a right handed system strictly in terms of the human body.

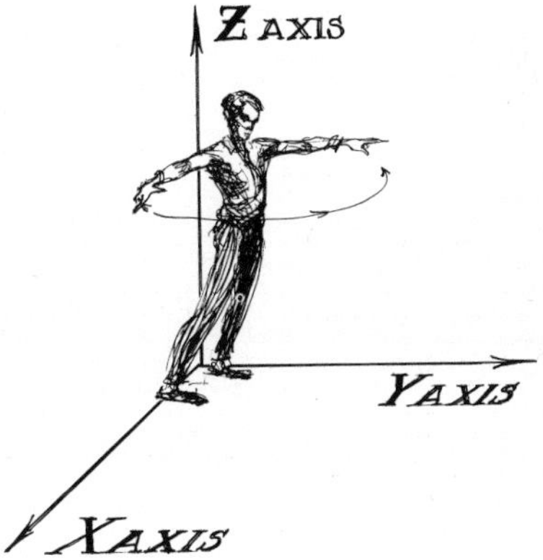

Consider a physical coordinate system whose axes are labeled x, y, and z.

This system is right handed if a man whose body lies along the z axis with his right arm pointing along the x axis obtains the y axis by rotating his right arm across the front of his body toward his left arm.

We can also describe this procedure by the right hand. Let the thumb of the right hand be along the axis corresponding to the z axis (the coordinate listed last in the triplet of numbers specifying a point, i.e., (x, y, z)). Initially the fingers of the open hand point along the x axis (the coordinate listed first in the triplet of numbers). The y axis is obtained by rotating the fingers toward the palm in an operation of closure.

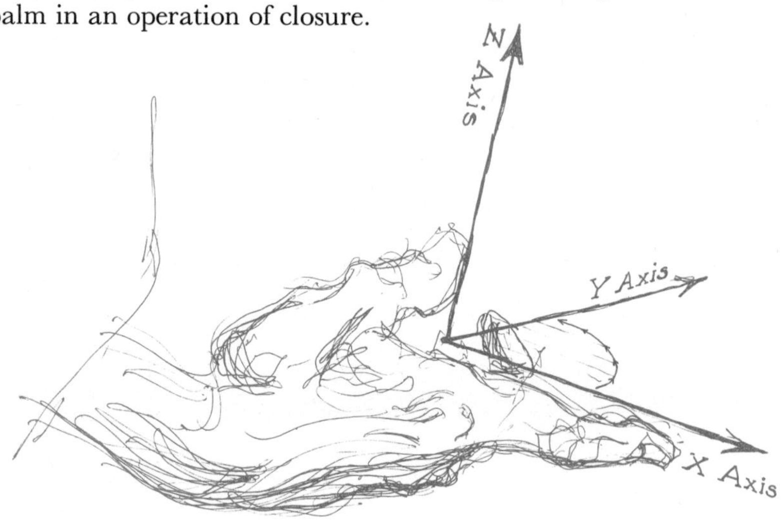

An inversion of any one of the axes of a right handed system gives a left handed system. By inversion of an axis we mean that the axis of the inverted system corresponds to the negative axis of the right handed system.

In the triplet of coordinates the order of writing the coordinate can indicate the handedness . The order can be changed in a cyclic order with changing the handedness. Our meaning of cyclic order can be shown by listing the "right handed" orders and the "left handed" orders.

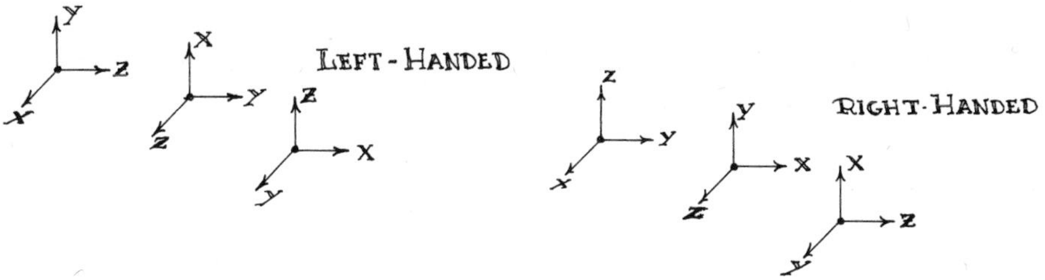

(x, z, y)		(x, y, z)	
(z, y, x)	**left handed**	(z, x, y)	**right handed**
(y, x, z)		(y, z, x)	

Although we shall not specifically deal with the order it is perhaps interesting to note this property here.

In addition we perceive that *there are only two orders possible.*

Below is shown a comparison of a "right handed" and a "left handed" system obtained by the inversion of the y axis.

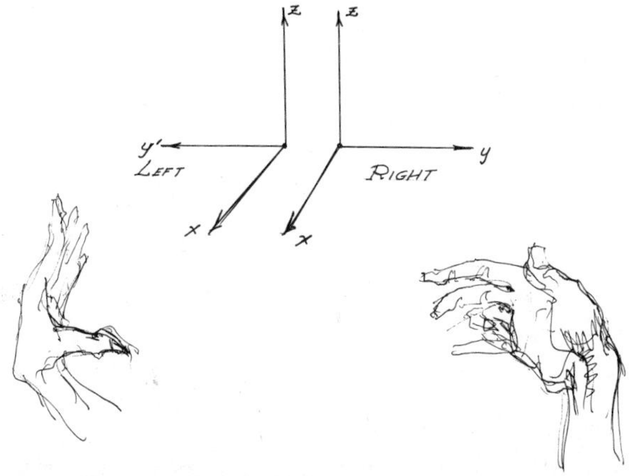

For many years scientists believed that the choice of coordinates could be taken arbitrarily. That is to say that the results of any physical experiment should not depend upon the left hand or right hand properties of the coordinates. Because of this arbitrariness the laws of physics were always set up to be independent of this particular property of the coordinates.

This property of a physical system is referred to as ***parity.***

In an attempt to describe this problem in terms of everyday observations; assume that the scientists on earth are able to make radio contact with a similar type of being somewhere in outer space, and further assume that the communications can be verbally understood.

Now hypothetically we find that this "being" in outer space has essentially the same make up as the "being" on earth, i.e., two arms, two hands, a head, two legs, a heart, etc. After much conversation we decide to correlate an experiment on earth and an experiment in outer space.

The experiment is that of setting up a magnetic field by winding a current coil. The instructions of the earth people for finding the direction of the magnetic field are as follows:

> "Take one current loop. Curl the fingers of the right hand in the direction of the current, and then the thumb of that hand points in the direction of the field."

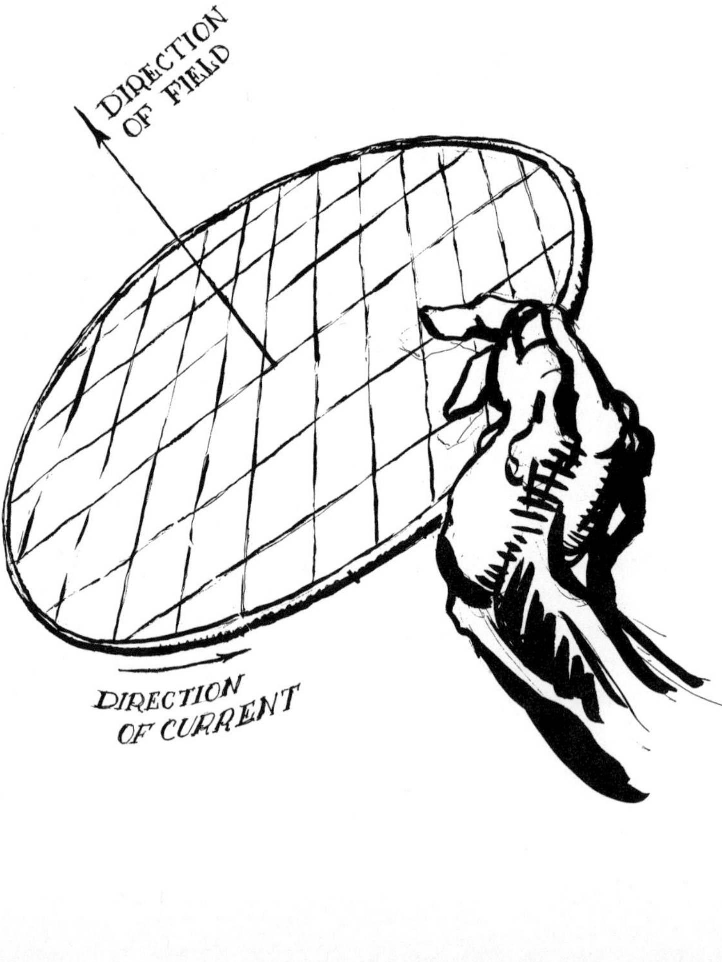

These instructions are conveyed to outer space, and at this stage we receive a question from our correspondent. This question provides a stumbling block. He asks,

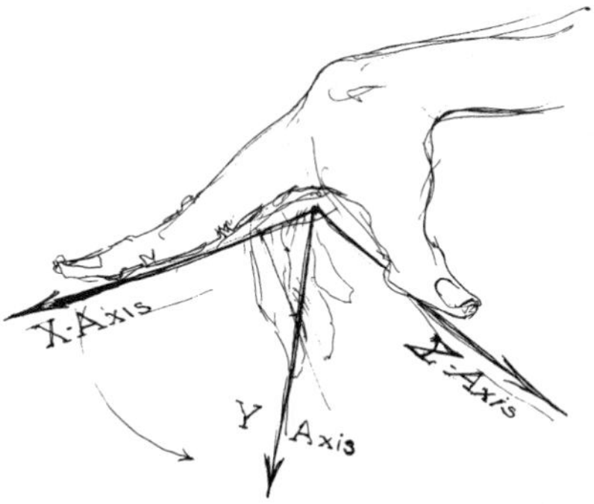

"Which of my two hands should I call the right hand?"

Our first impulse is to say that his right hand is on the side opposite his heart. Let us believe for the sake of argument that his reply is to the effect that his heart is in the center of his body midway between the shoulders. Therefore we cannot find any way to uniquely choose his right hand.

For this reason (keeping our special example of current loop and magnetic field) for many years scientists assumed that the results of experiments in magnetic fields would not uniquely determine the direction of the field. More explicitly let us say that if a radioactive source emitting electron is placed in a magnetic field it was expected that just as many electrons would be emitted in the direction of the field as would be emitted in the opposite direction.

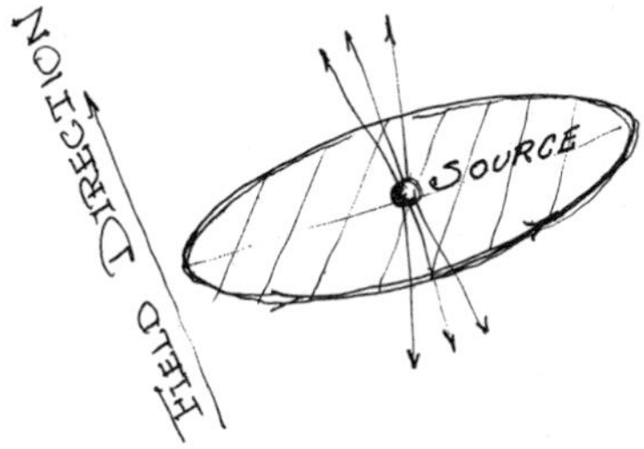

In 1956 because of questions concerning this postulate proposed by Dr. Frank Yang and Dr. T. D. Lee, experiments showed that certain natural phenomena did indeed uniquely describe the difference between a right and left handed coordinate system.

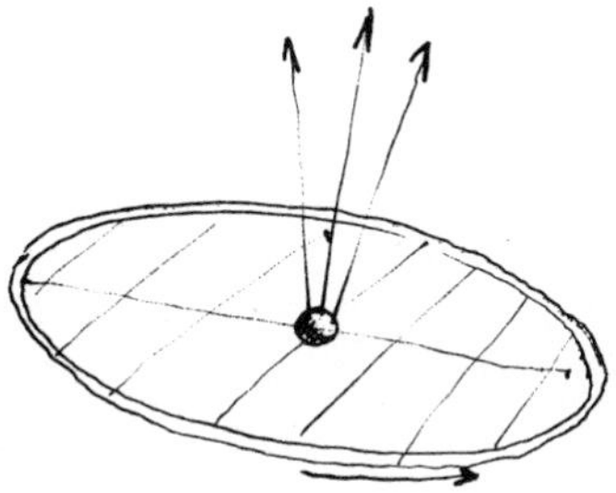

A casual description of the results is that in the presence of an external magnetic field certain electron emitting radioactive sources do emit more electrons in one direction than in another. These experiments are called the ***parity experiments.***

Thus to find the right handed system for the man in outer space we could ask him to repeat the ***parity experiments*** and after noting the direction corresponding to the emission of an excess number of electrons he would be able to uniquely assign a direction to the magnetic field.

Curvilinear Coordinates

In many problems of physical and mathematical interest the so-called "curvilinear coordinate systems" are the more convenient for a given problem. Since a point in physical space requires three numbers to define it relative to a specified coordinate system, it is possible to specify the point by three numbers other than the length measured along the x, y, z axes of E_3.

To introduce this possibility in the simplest case consider the two-dimensional space E_2. In addition to the doublet (x, y) which specifies a point P we can choose a polar coordinate system which provides us with a doublet consisting of the distance ρ of the point P from the origin and the angle ϕ which the line connecting the origin and P makes with the x axis. (In the *xy plane* of E_3 we shall also denote the radial distance to the projection of P by the coordinate ρ, and we shall denote the angle between ρ and the x axis by the coordinate angle ϕ).

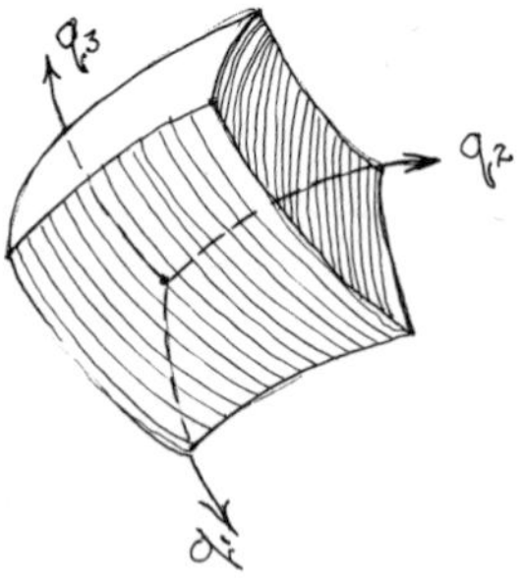

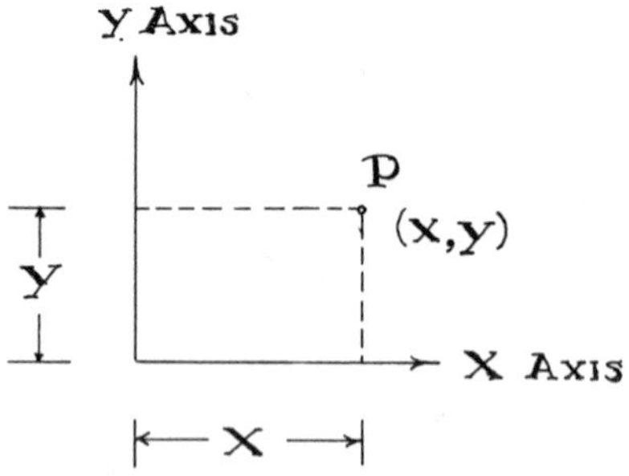

A comparison of coordinates is shown below.

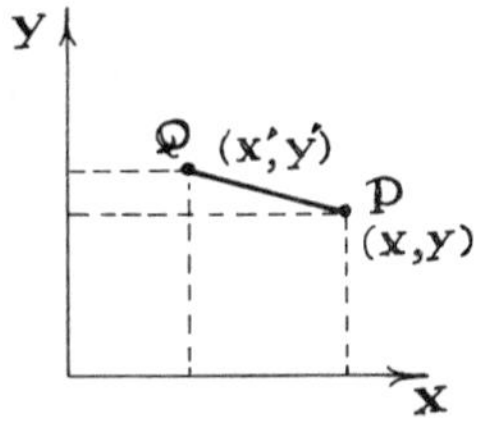

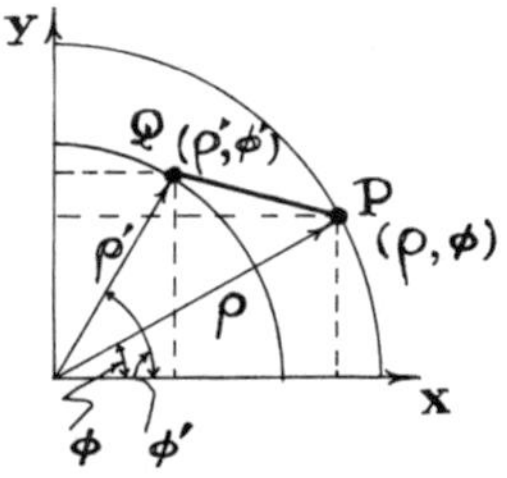

Other curvilinear systems are possible but need not be discussed here.

Even more interesting are the curvilinear systems of E_3. The two most important curvilinear systems are the ***cylindrical coordinate system*** and the ***spherical coordinate system.***

The cylindrical coordinate system takes as coordinates ρ and ϕ of the polar system in the xy plane *plus* the z coordinate of E_3. The point P is located by the intersection of three surfaces;

1 The cylindrical surface of radius ρ having its axis coincide with the z axis.

2 A plane containing the z axis and the point P.

3 A plane parallel to the xy plane at a distance z from the xy plane.

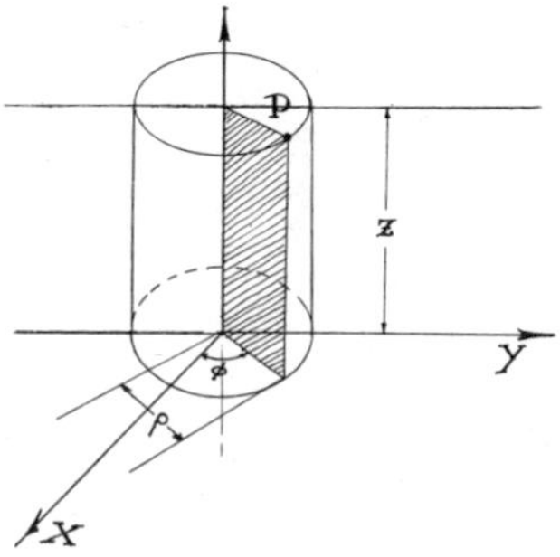

A point P in **the spherical system** is also located by two surfaces and a line which define the associated coordinates. These parameters are:

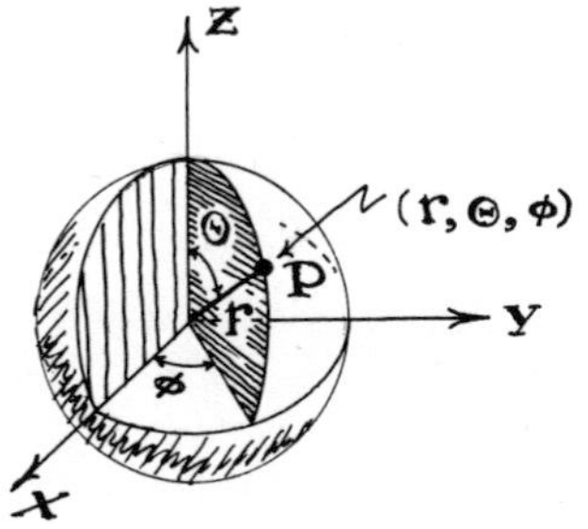

1 A spherical surface whose center lies at the origin and which passes through the point P. This spherical surface has a radius r, the distance between the origin and the point P ($r^2 = x^2 + y^2 + z^2$).

2 The straight line between the origin and P makes an angle θ with the z axis.

3 A plane containing the z axis and the point P. The line of intersection of this plane and the xy plane makes an angle ϕ with respect to the x axis.

There are various other curvilinear systems defined by ellipsoidal, hyperbolic surfaces, etc. However, our main intent is merely to indicate by the examples above the possibility of systems other than cartesian.

The Spherical Coordinates incidently are very useful in problems which deal with motion relative to the surface of the earth.

The Length Element

The functional dependence on the coordinate variables which the length element "d" takes in a curvilinear coordinate system can be obtained in two ways, both of which are equivalent.

The first method which can be used is that of geometrical construction. The second method is performed by first obtaining the relations between x, y, z and the variables of the curvilinear system and then substituting directly into the equation.

$$d^2 = (x - x')^2 + (y - y')^2 + (z - z')^2$$

To illustrate this we shall consider the distance d between a point P (x,y) and a point Q (x′,y′) in E_2 and in two dimensional polar coordinates with P at (ρ,ϕ) and Q at (ρ',ϕ').

In the diagram below the two coordinate systems are shown.

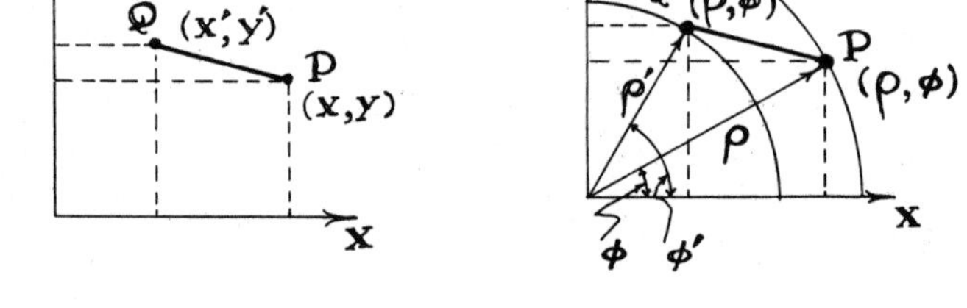

By axiom,

$$d^2 = (x - x')^2 + (y - y')^2$$

By the law of cosines.

$$d^2 = \rho^2 + \rho'^2 - 2\rho\rho' \cos(\phi' - \phi)$$

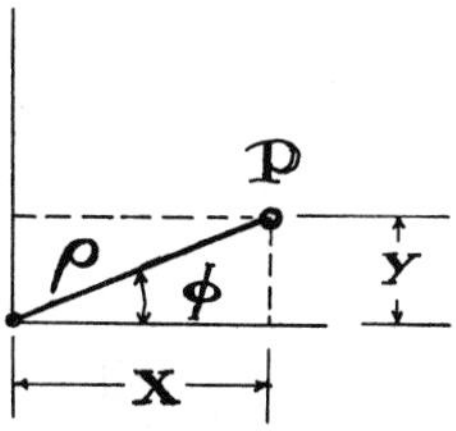

To illustrate the analytic technique we see that

$$x = \rho \cos\phi$$

and

$$y = \rho \sin\phi$$

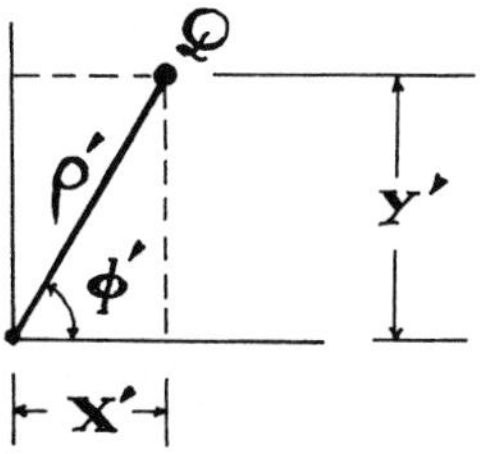

Also

$$x' = \rho' \cos \phi'$$

and

$$y' = \rho' \sin \phi'$$

Thus

$$(x - x')^2 = (\rho \cos\phi - \rho' \cos\phi')^2$$

and

$$(y - y')^2 = (\rho \sin\phi - \rho' \sin\phi')^2$$

Giving

$$d^2 = (x - x')^2 + (y - y')^2 = \rho^2 + \rho'^2 - 2\rho\rho' \cos(\phi' - \phi)$$

The procedure by which we obtain d^2 in the three dimensional systems is the same. The details of the development can be seen from the diagrams; the results are merely stated.

In cylindrical coordinates.

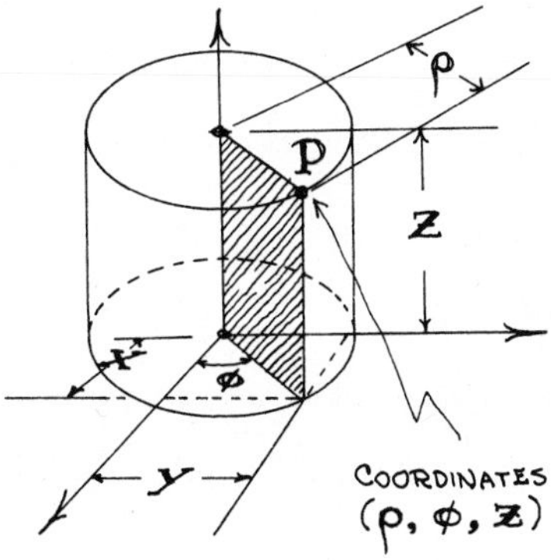

$$x = \rho \cos \phi \qquad x' = \rho' \cos \phi'$$
$$y = \rho \sin \phi \qquad y' = \rho' \sin \phi'$$
$$z = z \qquad z' = z'$$

and

$$d^2 = \rho^2 + \rho'^2 - 2\rho\rho' \cos(\phi' - \phi) + (z' - z)^2$$

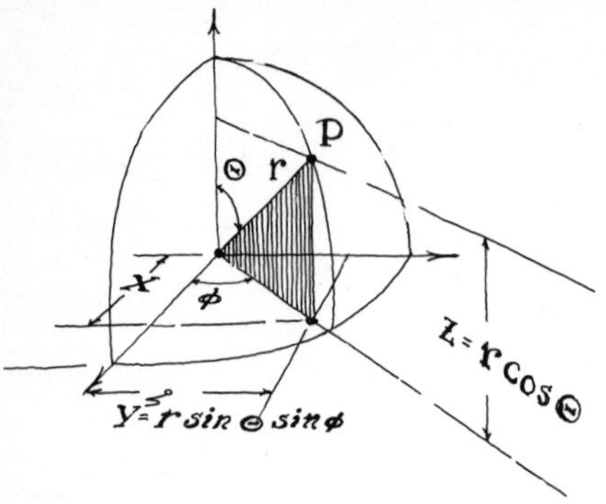

In spherical coordinates.

$$x = r\sin\theta\cos\phi \qquad x' = r'\sin\theta'\cos\phi'$$
$$y = r\sin\theta\sin\phi \qquad y' = r'\sin\theta'\sin\phi'$$
$$z = r\cos\theta \qquad z' = r'\cos\theta'$$

Thus

$$d^2 = (r\sin\theta\cos\phi - r'\sin\theta'\cos\phi')^2 + (r\sin\theta\sin\phi - r'\sin\theta'\sin\phi')^2 + (r\cos\theta - r'\cos\theta')^2$$

and

$$d^2 = r^2 + r'^2 - 2rr'\{\cos\theta\cos\theta' + \cos(\phi' - \phi)\sin\theta\sin\theta'\}$$

The reader may verify this result by expanding the squared terms.

The length element d can be approximated in a simple form when d is **much much** smaller than x and y.

Let Δx be a number much smaller than x or y, and let Δy be a number much smaller than y or x.

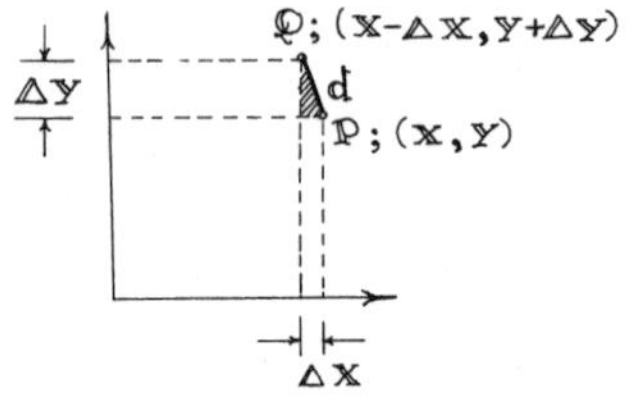

If Q is very close to P

$$x' = x \pm \Delta x$$

and $\quad y' = x \pm \Delta y$

then $\quad d^2 = (\Delta x)^2 + (\Delta y)^2$

For the two dimensional polar coordinates, let

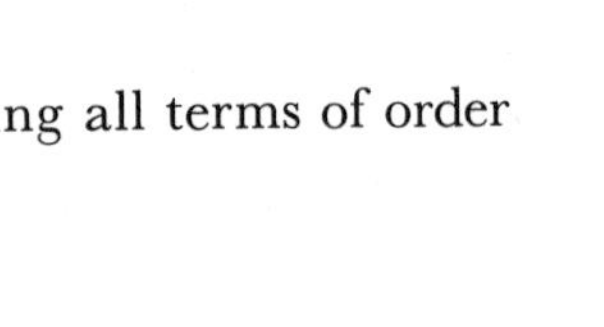

$$\rho' = \rho + \Delta\rho$$

and $\quad \phi' = \phi + \Delta\phi$

then

$$d^2 = \rho^2 + (\rho + \Delta\rho)^2 - 2\rho(\rho + \Delta\rho)\cos(\Delta\phi)$$

Expanding the squared terms and dropping all terms of order $\Delta\rho(\Delta\phi)^2$ and higher[5], we find that

$$d^2 \simeq (\Delta\rho)^2 + \rho^2(\Delta\phi)^2$$

[5] We must use the expansion of $\cos(\Delta\phi)$ in powers of $\Delta\phi$, i.e., $\cos(\Delta\phi) = 1 - \frac{1}{2!}(\Delta\phi)^2 + \frac{1}{4!}(\Delta\phi)^4 \text{ - - - - } (-1)^n \frac{1}{(2n)!}(\Delta\phi)^{2n}$

The Carousel Problem

Problems which are set up in curvilinear coordinates are of particular interest to us because our frames of reference on the surface of the earth are strictly curvilinear frames. Sitting in the class room we find that the assumption of a cartesian space is a sufficiently good approximation for many purposes.

Whenever the observer attempts to extend the cartesian approximation too far he finds anomalies which can only be explained by reverting to the true curvilinear system.

In order to gain some insight into these problems let us consider a two dimensional curvilinear system; in particular let us compare observations made by an observer standing on the earth with those observations made by an observer standing upon a merry-go-round. For simplicity we shall assume that the merry-go-round rotates with a constant angular velocity.

The laws of physics will be considered as true in the frame fixed to the surface of the earth (to a good approximation). *We shall find that certain phenomena which exhibit a particularly simple behavior in the earth's frame become quite complicated when described by the observer in the rotating frame.*

Suppose for instance that the man on the carousel drops a marble from the edge of the carousel.

The observer on the earth's surface sees the marble move off in a straight line tangent to the carousel at the point of dropping.

The observer on the merry-go-round however sees the marble curve radially outward and fall behind.

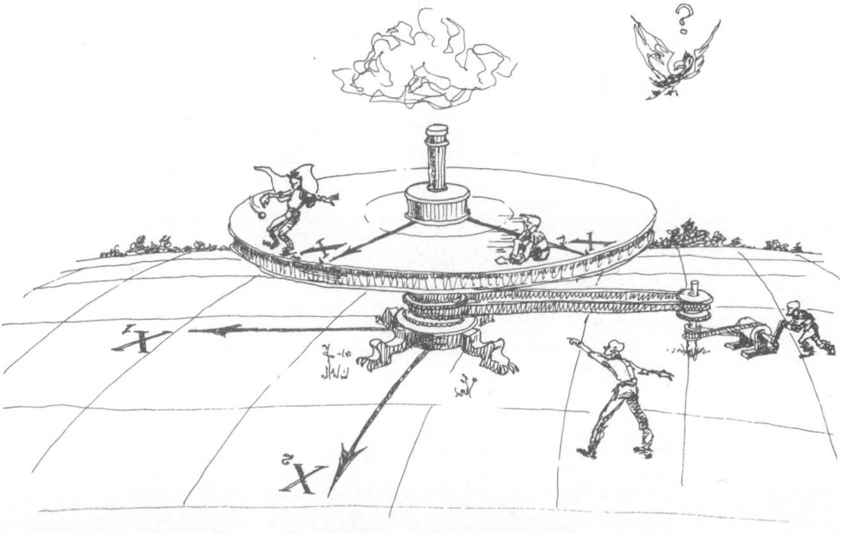

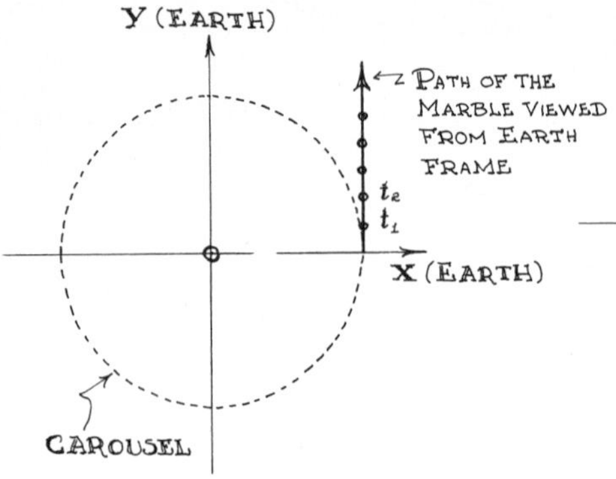

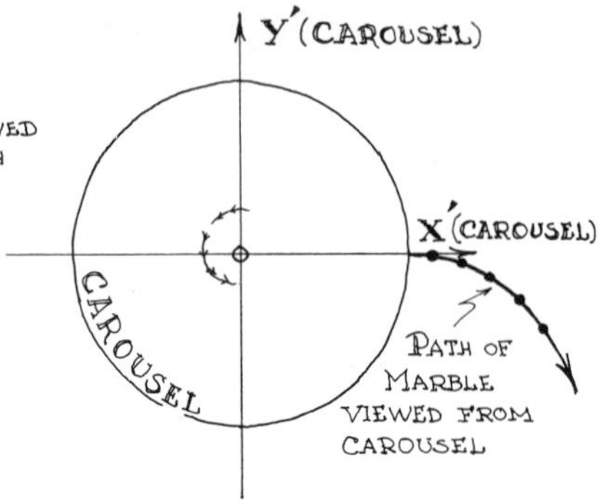

Initially the rotating observer sees the marble moving radially away from him. He interprets this as some fictitious force which acts upon all objects in his frame of reference. To the force he might give the name centrifugal, and the radial acceleration which he observes he might call centrifugal acceleration.

If the rotating observer is aware that the laws of physics hold in their given form in the stationary frame only he can then explain quite well the observations made from the carousel.

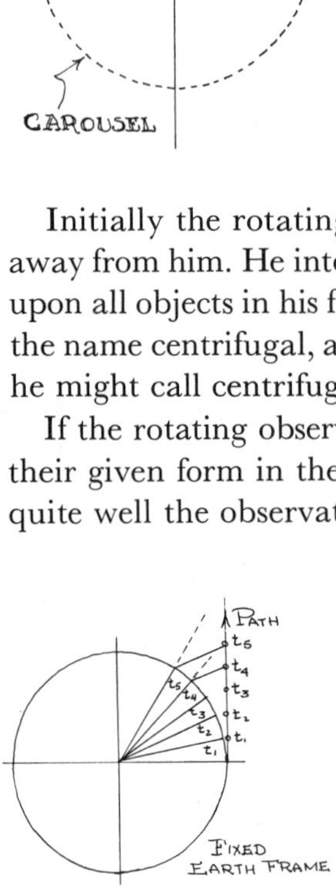

By relating positions of the marble along the straight line in the stationary frame with the predicted positions of his observation point at corresponding times in the rotating frame he can predict the path observed from the rotating frame.

It is sufficient to state here that if the straight line path of the marble in the stationary frame is transformed to the rotating curvilinear frame of the carousel; terms corresponding to the centrifugal and Coriolis accelerations appear quite naturally.

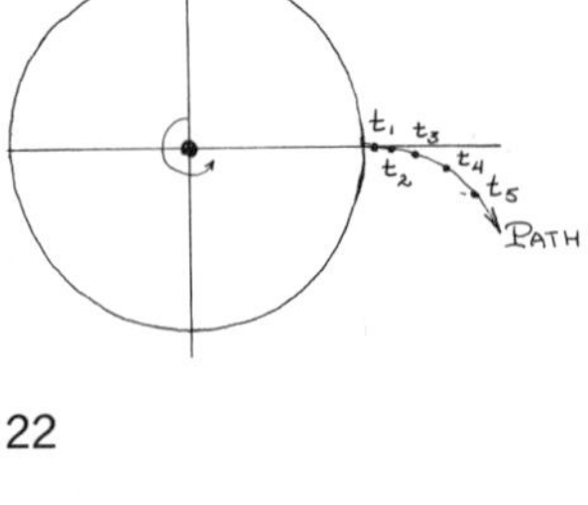

For the second problem, assume that the observer standing on the circumference of the carousel wishes to shoot an arrow into a post at the center.

In his first attempt the rotating observer aims directly at the center. He finds that the arrow curves away from the center in the direction of rotation of the carousel. This deviation he again attributes to a fictitious force which he calls the *Coriolis Force.*

To strike the center post the rotating observer finds that he must fire the arrow in a direction which compensates for the rotation of the carousel.

The explanation for this is quite simple when viewed from the stationary frame, at the instant of firing the arrow has two velocities.

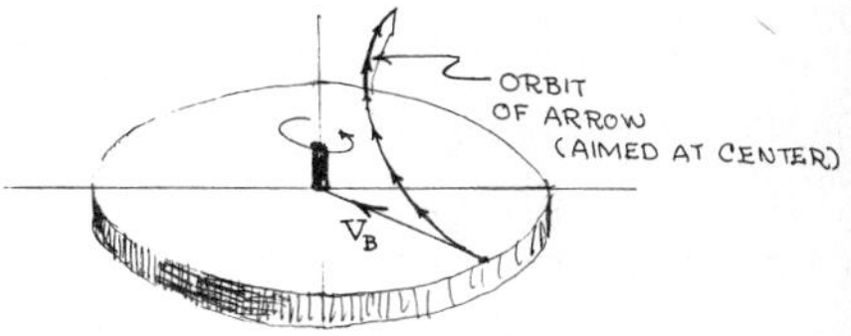

One velocity provided by the bow $\mathbf{V_B}$; the other velocity $\mathbf{V_t}$ arising from the motion of the point of firing with respect to the stationary system.

The velocity of the arrow relative to the earth is the vector sum of the velocities $\mathbf{V_B}$ and $\mathbf{V_t}$.

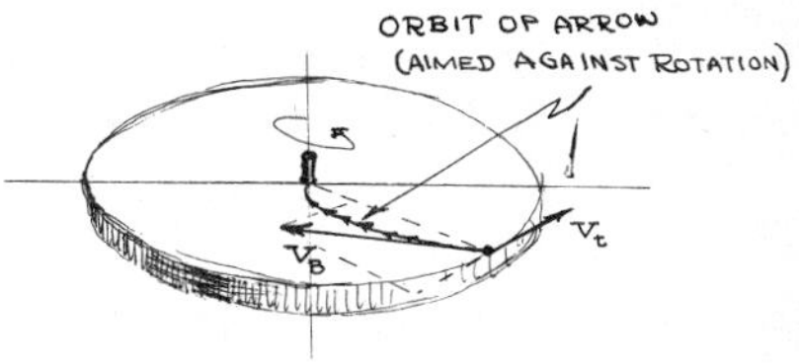

Thus to strike the center point this resultant velocity must be directed along the line joining the firing point and the center.

Again the path of the arrow once it has been fired is a straight line in the stationary system; while its path when viewed from the rotating frame is a complicated curve.

D. Coordinate Transformations

In order to introduce the concept of changes in the coordinates or coordinate transformations in an elementary manner, we shall confine our discussion to a two dimensional space.

Of the various changes or transformations which can be performed, only two will be of interest to us at this point; ***translation of the origin*** and ***rotation about the origin.***

Translations

The simplest of all transformations is the *operation* of translation. Consider the space E_2 shown in the diagram to follow.

On the left is the original coordinate system O with the coordinates of a point P being x and y. To the right of this diagram is the coordinate system after the origin has been shifted to O′ (a distance A in the direction of the x axis and a distance B in the direction of the y axis). Rotation is specifically excluded and thus the new coordinate axes are parallel to the old.

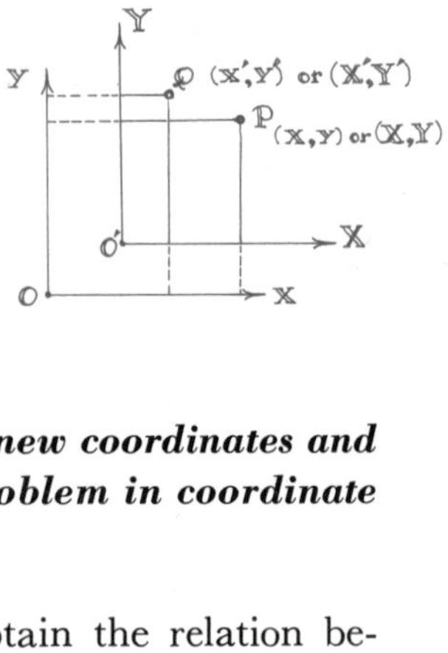

Obtaining algebraic relations between the new coordinates and the old coordinates is the fundamental problem in coordinate transformations.

From the construction we can readily obtain the relation between (x,y) and (X,Y).

$$X = x - A, \text{ and } Y = y - B$$

It is important to find in this elementary case that the operation of translation does *not* change the distance "d" between the two points P and Q.

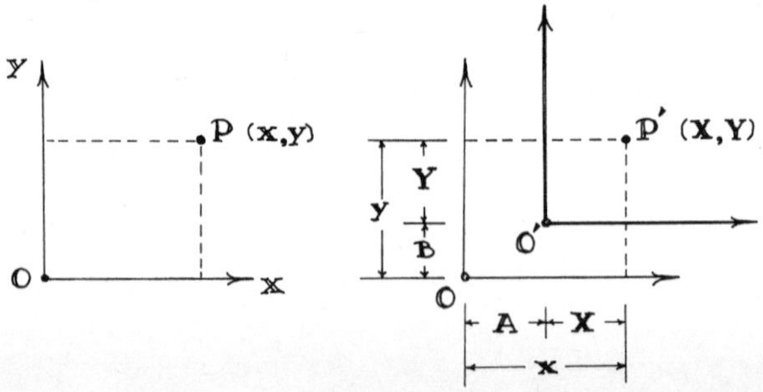

In more elegant language we say that the distance "d" is ***invariant*** (unchanged in magnitude) under the operation of translation.

$$d^2 = (x - x')^2 + (y - y')^2 = (X - X')^2 + (Y - Y')^2$$

Rotations in E_2

We shall approach the problem of rotating the coordinate system about the origin first as a purely geometric problem. Once we have obtained the relations between the coordinates after rotation and the coordinates previous to the rotation, we shall show how the results of the geometric problem can be set up as a problem involving a ***matrix operation.*** In order to facilitate the transition from one set of expressions to the other it will prove convenient to *change* our notation slightly.

WARNING

a simple change in notation

let x be written as x_1

and y be written as x_2

The subscripts 1 on x_1 and 2 on x_2 serve to distinguish the different coordinates in E_2 (namely the coordinates which formerly were denoted by x and y). In other words the subscripts 1 and 2 are labels only (*not* multipliers).

Now rotate the coordinate system about the origin O through an angle α.

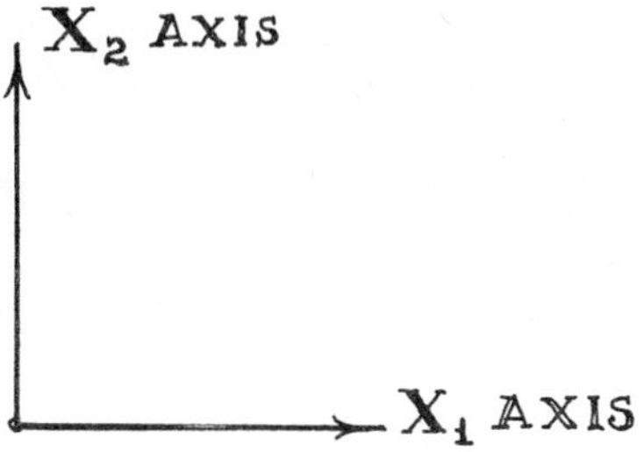

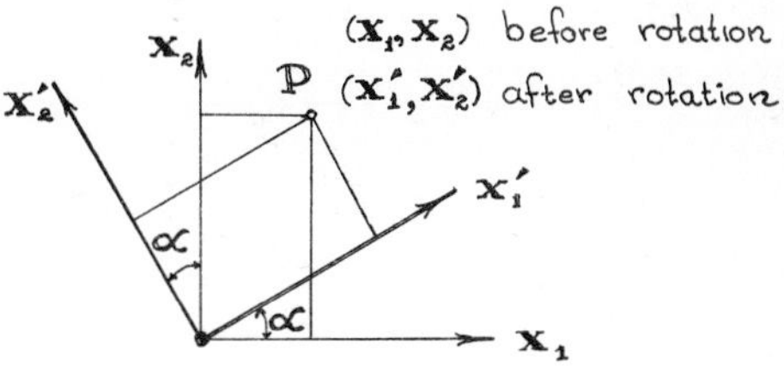

The new coordinates of the point P are x_1' and x_2'.

By the elementary procedures of trigonometry we can obtain the relations between x_1, x_2 and x_1' and x_2'.

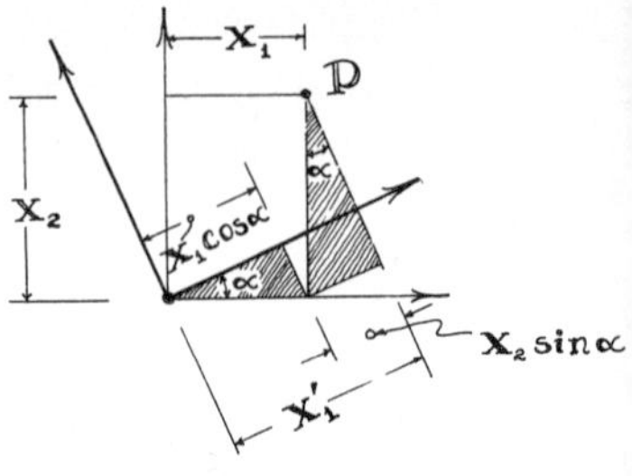

$$x_1' = (\cos\alpha)x_1 + (\sin\alpha)x_2$$

In the same manner we can find x_2' by triangulation.

$$x_2' = (-\sin\alpha)x_1 + (\cos\alpha)x_2$$

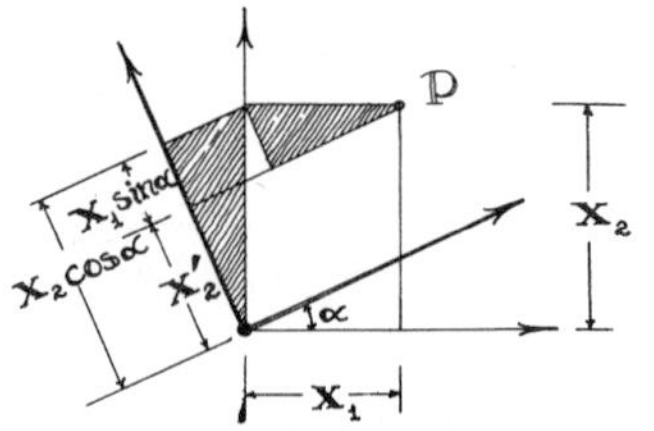

The coordinates before rotation, x_1 and x_2, are related to the coordinates after rotation, x_1' and x_2', by the **linear algebraic equations** above. This is called a **linear transformation** of the coordinates x_1 and x_2 to the coordinates x_1' and x_2'.

To illustrate this transformation of coordinates regard the following simple problem.

Assume that there is a two dimensional cartesian coordinate system laid out upon the surface of the earth with the origin at O.

Now we construct a circular merry-go-round with its axis of rotation at O. Coordinate axes x_1' and x_2' are painted upon the floor of the merry-go-round.

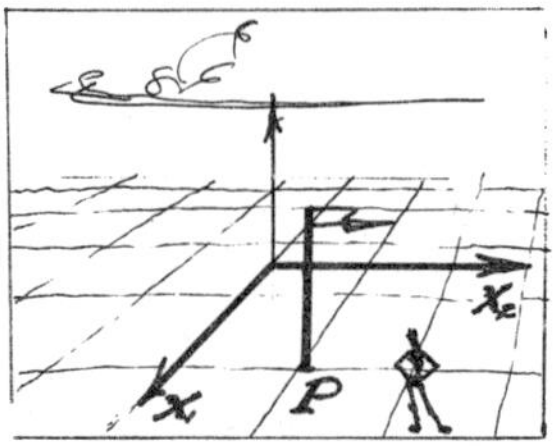

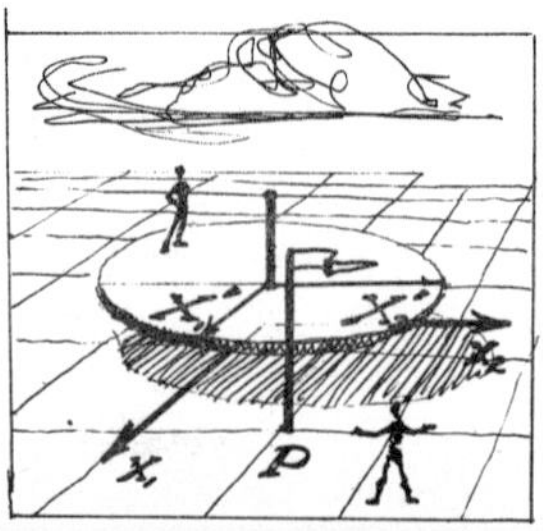

As long as the axes x_1' and x_1 are aligned the coordinates of a given point have the same value in (x_1, x_2) as in (x_1', x_2'). This situation obviously corresponds to a **rotation of zero degrees,** i.e., $\alpha = O$.

Thus if

$$x_1' = (\cos\alpha)x_1 + (\sin\alpha)x_2$$

and

$$x_2' = (-\sin\alpha)x_1 + (\cos\alpha)x_2$$

Then for the **particular case** $\alpha = O$

$$\left.\begin{aligned} x_1' &= x_1 \\ x_2' &= x_2. \end{aligned}\right\} \text{ Axes aligned.}$$

From the condition of alignment let us now compare the location of a point P with coordinates (x_1, x_2) in the system fixed upon the earth with the observed coordinates of the point P relative to the axes fixed upon the merry-go-round when the merry-go-round rotates through an angle α of 30° counterclockwise.

In this situation $\alpha = 30°$ and

$$x_1' = (\cos 30°)\, x_1 + (\sin 30°)\, x_2 = \left(\frac{\sqrt{3}}{2}\right) x_1 + \left(\frac{1}{2}\right) x_2$$

and

$$x_2' = (-\sin 30°)\, x_1 + (\cos 30°)\, x_2 = \left(-\frac{1}{2}\right) x_1 + \left(\frac{\sqrt{3}}{2}\right) x_2$$

If we are now told that

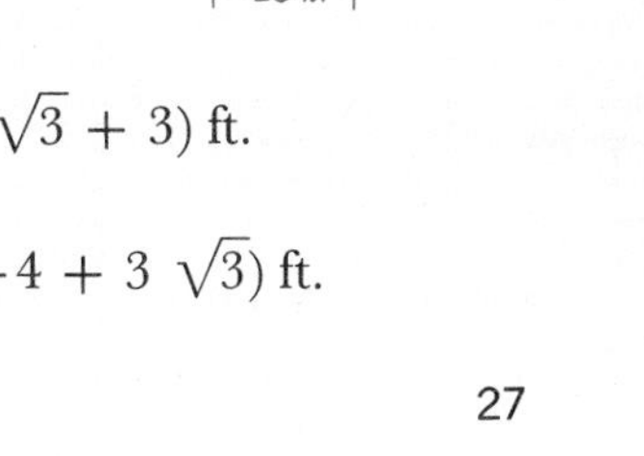

$$x_1 = 40 \text{ ft.}$$

and

$$x_2 = 30 \text{ ft.}$$

then

$$x_1' = \frac{\sqrt{3}}{2}(40) + \frac{1}{2}(30) = 5\,(4\,\sqrt{3} + 3) \text{ ft.}$$

$$x_2' = -\frac{1}{2}(40) + \frac{\sqrt{3}}{2}(30) = 5\,(-4 + 3\,\sqrt{3}) \text{ ft.}$$

These two ***algebraic expressions can be written in abbreviated forms*** and ***in terms of matrices.***

The field of matrices and linear transformations is naturally much broader than our example of the rotation of a coordinate system of two dimensions. However, once the use of matrices is understood in this simple problem the extension of the method to the more general problem is quite straight forward.

Introduction of New Symbols

let

$$x_1' = (\cos\alpha)\, x_1 + (\sin\alpha)\, x_2 = S_{11}\, x_1 + S_{12}\, x_2$$

and

$$x_2' = (-\sin\alpha)\, x_1 + (\cos\alpha)\, x_2 = S_{21}\, x_1 + S_{22}\, x_2$$

thus

$$S_{11} = \cos\alpha, \qquad S_{12} = \sin\alpha$$
$$S_{21} = -\sin\alpha, \qquad S_{22} = \cos\alpha$$

We now introduce a further contraction of the notation presented above. We note that each equation is linear and has a general form.

$$x_j' = \sum_{k=1}^{2} S_{jk}\, x_k$$

The indices (or labels j and k) can be either 1 or 2. For instance $j = 1$, and we sum over k we obtain the first equation. We further find that in the case that $j = 2$, we obtain the second equation, remembering that S_{jk} is defined above,

$$x_j' = \sum_{k=1}^{k=2} S_{jk}\, x_k = S_{j1}\, x_1 + S_{j2}\, x_2$$

The summation sign Σ can be eliminated by the EINSTEIN **summation convention.** This convention *merely* drops the summation sign with the understanding that the repeated index k in $S_{jk}\,x_k$ *always* demands a sum over all of the possible values of k.

$$x_j' = \sum_{k=1}^{2} S_{jk}\,x_k = S_{jk}\,x_k$$

Take an example: Let j = 1 then

$$x_1' = S_{1k}\,x_k = S_{11}\,x_1 + S_{12}\,x_2$$

Note that all of this does *not* consist of introducing anything new. This procedure only makes the equations more concise and allows us to put more information into a *smaller space.* In such a form, however, the technique can be readily extended.

To reiterate:

$x_j' = S_{jk}\,x_k$ implies in our simple problem,

$$x_1' = (\cos\alpha)\,x_1 + (\sin\alpha)\,x_2$$
$$x_2' = (-\sin\alpha)\,x_1 + (\cos\alpha)\,x_2$$

These **algebraic equations** can also be written in the form of **matrices.**

E. Matrices

The characteristics of a set of **linear algebraic equations** lead to a particularly convenient and powerful representation in terms of **matrices.**

A matrix is a rectangular array of mathematical objects. In our particular problem[6] the matrices will be

square,

and the objects in the array will be

BECAUSE OF A GOOD FORTUNE

$$g_m = \sum_n A_{mn}\,f_n$$

[6] In general the matrix can be rectangular; it need not be restricted the square way.

numbers;

i.e., the coefficients of the algebraic equations.

The square array

$$A = \left(\begin{array}{c|c} A_{11} & A_{12} \\ \hline A_{21} & A_{22} \end{array}\right)$$

is a **square matrix.** The 2 x 2 matrix above has been written in a general form having four components A_{11}, A_{12}, A_{21}, and A_{22}. These four components will have numerical values which are specified by *individual problems.* It is customary to perform the general discussion of matrices with the symbols A_{lm}.

For instance the array

$$\left(\begin{array}{c|c} 6 & -1 \\ \hline 0 & 57 \end{array}\right)$$

is a matrix. In this specific example the symbols A_{nm} have been assigned specific numerical values; namely

$$A_{11} = 6 \qquad A_{12} = -1$$
$$A_{21} = 0 \qquad A_{22} = 57$$

This assignment has no general significance except to illustrate the form in which one might find a numerical matrix.

Notice that while introducing this representation to the reader we have **"sectioned off"** the various positions in the array by dashed lines. *Ordinarily* this is NOT done and the matrix appears in the following fashion:

$$\begin{pmatrix} 6 & -1 \\ 0 & 57 \end{pmatrix}$$

The technique of sectioning the matrix will be continued until the reader has become familiar with the representation.

Before proceeding to the laws for matrix manipulation it should be recognized that the use of the 2 x 2 square array is not intended in any manner to limit the operations discussed to 2 x 2 matrices alone. Our remarks concerning addition and multiplication, etc. will apply equally well to arrays of higher order.

An example of a 4 x 4 array is shown below.

$$A = \left(\begin{array}{c|c|c|c} A_{11} & A_{12} & A_{13} & A_{14} \\ \hline A_{21} & A_{22} & A_{23} & A_{24} \\ \hline A_{31} & A_{32} & A_{33} & A_{34} \\ \hline A_{41} & A_{42} & A_{43} & A_{44} \end{array}\right)$$

Notice carefully the indices of the components of these matrices. Any one of the components is denoted by the general symbol,

$$A_{nm}$$

where the indices n and m can have any one of the four values 1, 2, 3, or 4.

Using A_{nm} as an example note that the leading index "n" indicates the row in which the component falls while the trailing index "m" specifies the column to which the component belongs.

Thus the component A_{32} is the component belonging to the **third row and** the **second column.**

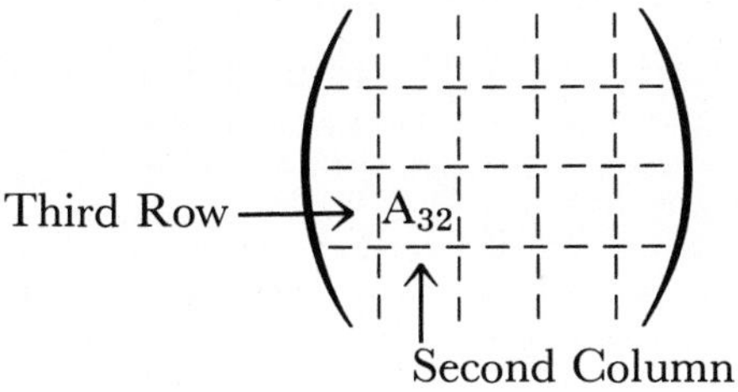

When two matrices are added the components of the resultant matrix are made up of the sums of the corresponding elements of the original matrix; i.e., $(\mathbb{A} + \mathbb{B})_{mn} = A_{mn} + B_{mn}$.

The addition of matrices obeys the

Commutative Law

$$\mathbb{A} + \mathbb{B} = \mathbb{B} + \mathbb{A}$$

The Associative Law

$$(\mathbb{A} + \mathbb{B}) + \mathbb{C} = \mathbb{A} + (\mathbb{B} + \mathbb{C})$$

and The Distributive Law

$$n(\mathbb{A} + \mathbb{B}) = n\mathbb{A} + n\mathbb{B}$$

The multiplication of a matrix by a scalar quantity n is accomplished by multiplying each component by the scalar in question. As an example we can write

$$n\mathbb{A} = n\begin{pmatrix} A_{11} & A_{12} \\ A_{21} & A_{22} \end{pmatrix} = \begin{pmatrix} nA_{11} & nA_{12} \\ nA_{21} & nA_{22} \end{pmatrix}$$

These properties of addition are stressed because the multiplication of matrices does *not* obey the commutative law.

The multiplication operation in the case of matrices obeys The Associative Law;

$$(\mathbb{A} \cdot \mathbb{B}) \cdot \mathbb{C} = \mathbb{A} \cdot (\mathbb{B} \cdot \mathbb{C})$$

The multiplication of matrices does not necessarily obey the **commutative law;**[7]

$$\mathbb{A} \cdot \mathbb{B} \neq \mathbb{B} \cdot \mathbb{A}$$

The product of two matrices can in very special cases commute, however in general commutation of matrix products is not expected.

[7] The symbol $\neq$ means **"does not equal."** In this case it is intended to mean not necessarily equal.

Before proceeding further with the details of taking products of matrices we should mention briefly a few of the characteristic matrices with which we shall be dealing.

The **transpose** of a matrix is obtained by interchanging ROWS and COLUMNS. Consider the matrix $\mathbb{A}$,

$$\mathbb{A} = \begin{pmatrix} A_{11} & A_{12} \\ A_{21} & A_{22} \end{pmatrix}$$

The **transpose** of $\mathbb{A}$ is written $\tilde{\mathbb{A}}$ with elements $\tilde{A}_{nm}$;

$$\tilde{\mathbb{A}} = \begin{pmatrix} \tilde{A}_{11} & \tilde{A}_{12} \\ \tilde{A}_{21} & \tilde{A}_{22} \end{pmatrix} = \begin{pmatrix} A_{11} & A_{21} \\ A_{12} & A_{22} \end{pmatrix}$$

In the illustration above the relation between the elements A_{nm} of the original matrix $\mathbb{A}$ and the elements $\tilde{A}_{jk}$ of the transposed matrix $\tilde{\mathbb{A}}$ is seen to be

$$\tilde{A}_{mn} = A_{nm}$$

It is apparent that the interchange of rows and columns corresponds to an interchange of indices on the components.

The **magnitude** of a square matrix is defined as the **determinant** of that matrix; thus

$$\text{Magnitude of } \mathbb{A} = |\mathbb{A}| = \text{Det } \mathbb{A}$$

The inverse of a matrix can be obtained if the magnitude (i.e., Det $\mathbb{A}$) is **non-zero.**

After the discussion of matrix multiplication the methods for finding the elements of the inverse matrix will be discussed.

One other form will prove extremely useful in future discussions; this matrix is **the unit matrix.**[8] The unit matrix is a square array having the value plus one for all diagonal elements and having the value zero for all off diagonal elements.

The elements of the unit matrix have a special symbol called the **Kronecker delta.**

[8] This matrix is sometimes called the **identity matrix.**

Kronecker delta $= \delta_{jm}$
$= 1$ when $j = m$
$= O$ when $j \neq m$

The symbol which we shall adopt for **unit matrix** is $\mathbb{I}$.

Thus the elements of $\mathbb{I}$ are δ_{jm}.

An example of a 3 x 3 unit matrix is shown below

$$\mathbb{I} = \begin{pmatrix} 1 & O & O \\ O & 1 & O \\ O & O & 1 \end{pmatrix}$$

Later as an exercise the reader should demonstrate that the unit matrix is the Identity Operator which is defined by the equation.

$$\mathbb{I} \cdot \mathbb{A} = \mathbb{A}$$

In other words the application of the unit matrix to a matrix $\mathbb{A}$ gives the same matrix $\mathbb{A}$.

Using the linear algebraic equations which define the transformation of (x_1, x_2) to (x_1', x_2') and using the square array representation for the matrix $\mathbb{S}$, we define the rules by which we multiply a *column matrix* (or vector) by a *square matrix.*

If a point P is defined by the coordinates x_1 and x_2 these coordinates can be set forth in an ordered column called a **column matrix** (or vector).

The column matrix locating the point P is represented by the symbol **r** and is written[9]

$$\mathbf{r} = \begin{bmatrix} x_1 \\ x_2 \end{bmatrix}$$

[9] The symbol **r** will be used to represent a vector in the chapter concerning vectors. The column matrix transforms like a vector and therefore will be referred to as such.

We have already represented $\mathbb{S}$ by the matrix

$$\mathbb{S} = \begin{pmatrix} S_{11} & S_{12} \\ S_{21} & S_{22} \end{pmatrix}$$

We multiply a row in the square matrix by the single column of the column matrix (or vector) to obtain a given element of the resulting column matrix.

Our two linear algebraic relations representing the rotation were

$$x_1' = S_{11}\, x_1 + S_{12}\, x_2$$
$$x_2' = S_{21}\, x_1 + S_{22}\, x_2$$

In terms of matrices we write this as

$$\begin{bmatrix} x_1' \\ x_2' \end{bmatrix} = \begin{pmatrix} S_{11} & S_{12} \\ S_{21} & S_{22} \end{pmatrix} \begin{bmatrix} x_1 \\ x_2 \end{bmatrix} = \begin{bmatrix} S_{11}\, x_1 + S_{12}\, x_2 \\ \hline S_{21}\, x_1 + S_{22}\, x_2 \end{bmatrix}$$

One observes that the product of a *column matrix* and a *square matrix* gives another column matrix.

Since the two column matrices above are equal their elements are equal.

$$\begin{bmatrix} x_1' \\ \hline x_2' \end{bmatrix} = \begin{bmatrix} S_{11}\, x_1 + S_{12}\, x_2 \\ \hline S_{12}\, x_1 + S_{22}\, x_2 \end{bmatrix}$$

i.e.,

$$x_j' = S_{j1}\, x_1 + S_{j2}\, x_2 = \sum_{k=1}^{k=2} S_{jk}\, x_k$$

To review this in terms of the previous carousel problem we again use a rotation of 30°.

The vector in the stationary system is $\begin{bmatrix} x_1 \\ x_2 \end{bmatrix} = \mathbf{r}$

The vector as seen from the merry-go-round is

$$\begin{bmatrix} x_1' \\ x_2' \end{bmatrix} = \mathbf{r}'$$

The operation of rotation through 30° is represented by the matrix

$$\begin{pmatrix} .866 & .500 \\ -.500 & .866 \end{pmatrix} = \mathbb{S}(30°)$$

The relation between the two sets of coordinates is provided by the equation,

$$\begin{bmatrix} x_1' \\ x_2' \end{bmatrix} = \begin{pmatrix} .866 & .500 \\ -.500 & .866 \end{pmatrix} \begin{bmatrix} x_1 \\ x_2 \end{bmatrix}$$

The student should carry out the matrix multiplication and show that the same result is obtained as before, i.e.,

$$x_1' = (.866)\,x_1 + (.500)\,x_2$$
$$x_2' = (-.500)\,x_1 + (.866)\,x_1$$

Matrix Multiplication

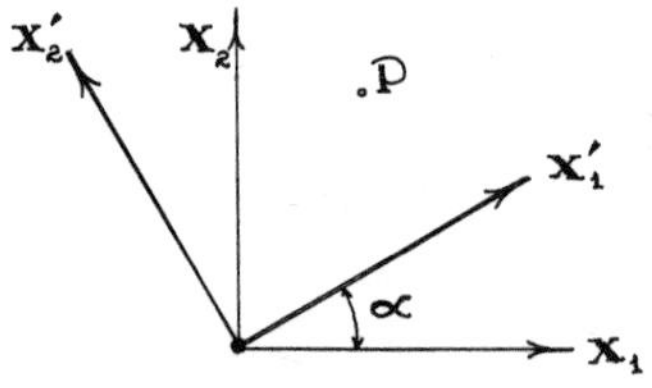

A single rotation of the coordinates x_1, x_2 to the set x_1', x_2' through an angle α has been shown to be represented by the equation

$$\mathbf{r}' = \mathbb{S}(\alpha) \cdot \mathbf{r}$$

where $S(\alpha)_{11} = \cos\alpha$, etc.

The parentheses α mean that $\mathbb{S}(\alpha)$ is the rotation matrix corresponding to a rotation through an angle α in a counterclockwise direction.

Consider now a second rotation of the system through an angle β taking the coordinates (x_1', x_2') to the coordinates (x_1'', x_2'').

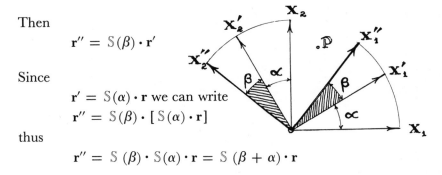

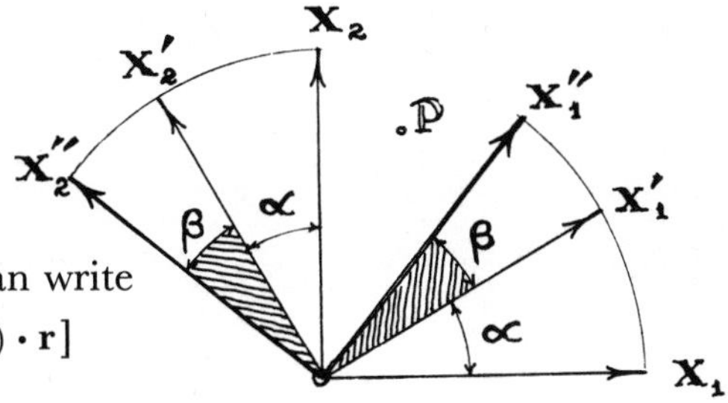

Then

$$r'' = \mathbb{S}(\beta) \cdot r'$$

Since

$$r' = \mathbb{S}(\alpha) \cdot \mathbf{r} \text{ we can write}$$
$$r'' = \mathbb{S}(\beta) \cdot [\mathbb{S}(\alpha) \cdot \mathbf{r}]$$

thus

$$r'' = \mathbb{S}\,(\beta) \cdot \mathbb{S}(\alpha) \cdot \mathbf{r} = \mathbb{S}\,(\beta + \alpha) \cdot \mathbf{r}$$

The relations between $\mathbb{S}(\beta) \cdot \mathbb{S}(\alpha)$ and $\mathbb{S}(\beta + \alpha)$ can be worked out quite simply by writing down the linear algebraic equations representing these transformations.

$r' = \mathbb{S}(\alpha) \cdot \mathbf{r}$ implies the two equations;

$$x_1' = (\cos\alpha)\,x_1 + (\sin\alpha)\,x_2$$

and

$$x_2' = (-\sin\alpha)\,x_1 + (\cos\alpha)\,x_2$$

The second rotation

$r'' = \mathbb{S}(\beta) \cdot r'$ indicates the following relations;

$$x_1'' = (\cos\beta)\,x_1' + (\sin\beta)\,x_2{}'$$

and

$$x_2'' = (-\sin\beta)\,x_1' + (\cos\beta)\,x_2'$$

By substituting the expansions of x_1' and x_2' into these last relations the transformation $\mathbb{S}(\beta + \alpha)$ is obtained.

$$x_1'' = (\cos\beta)\{(\cos\alpha)x_1 + (\sin\alpha)x_2\} + (\sin\beta)\{(-\sin\alpha)x_1 + (\cos\alpha)x_2\}$$

$$x_2'' = (-\sin\beta)\{(\cos\alpha)x_1 + (\sin\alpha)x_2\} + (\cos\beta)\{(-\sin\alpha)x_1 + (\cos\alpha)x_2\}$$

or

$$x_1'' = [\cos\beta\cos\alpha - \sin\beta\sin\alpha]\,x_1 + [\cos\beta\sin\alpha + \sin\beta\cos\alpha]\,x_2$$

$$x_2'' = [-\sin\beta\cos\alpha - \sin\alpha\cos\beta]\,x_1 + [-\sin\beta\sin\alpha + \cos\beta\cos\alpha]\,x_2$$

These relations simplify to the forms

$$x_1'' = (\cos(\beta+\alpha))x_1 + (\sin(\beta+\alpha))x_2$$

$$x_2'' = (-\sin(\beta+\alpha))x_1 + (\cos(\beta+\alpha))x_2$$

This demonstrates that $\mathbb{S}(\beta)\cdot\mathbb{S}(\alpha)$ is indeed $\mathbb{S}(\beta+\alpha)$ and if we examine the forms closely we find that the element $S(\beta+\alpha)_{nm}$ is the product of the nth ***row*** in $\mathbb{S}(\beta)$ times the mth **column** in $\mathbb{S}(\alpha)$.

In other words

$$S(\beta+\alpha)_{nm} = \sum_{k=1}^{2} S(\beta)_{nk}\,S(\alpha)_{km}$$

$$= S(\beta)_{n1}\,S(\alpha)_{1m} + S(\beta)_{n2}\,S(\alpha)_{2m}$$

To illustrate this regard the S_{11} element of $\mathbb{S}(\beta+\alpha)$

$$\mathbb{S}(\beta+\alpha) = \mathbb{S}(\beta)\cdot\mathbb{S}(\alpha) = \begin{pmatrix} \cos\beta & \sin\beta \\ -\sin\beta & \cos\beta \end{pmatrix}\begin{pmatrix} \cos\alpha & \sin\alpha \\ -\sin\alpha & \cos\alpha \end{pmatrix}$$

$$= \begin{pmatrix} \cos(\beta+\alpha) & \sin(\beta+\alpha) \\ -\sin(\beta+\alpha) & \cos(\beta+\alpha) \end{pmatrix}$$

The shaded areas indicate one particular **row-column** multiplication.

Here,

$$S(\beta + \alpha)_{11} = (S(\beta) \cdot S(\alpha))_{11} = S(\beta)_{11}\, S(\alpha)_{11} + S(\beta)_{12}\, S(\alpha)_{21}$$

$$= \cos\beta\cos\alpha - \sin\beta\sin\alpha = \cos(\beta + \alpha)$$

Once we have demonstrated the necessary multiplication rules required to provide consistency in the rotation through two successive angles, the general problem of multiplication can be discussed.

The multiplication of one matrix by another can be considered (for the purposes of introduction) as the multiplication of **an ordered array** of **column vectors** by a matrix. The result must of course be an **ordered array** of **column vectors** which is merely the **resultant matrix.**

To illustrate this let us compute the **product** $\mathbb{A} \cdot \mathbb{B}$ written as

$$\mathbb{A} \cdot \mathbb{B} = \begin{pmatrix} A_{11} & A_{12} \\ A_{21} & A_{22} \end{pmatrix} \begin{bmatrix} B_{11} \\ B_{21} \end{bmatrix} \begin{bmatrix} B_{12} \\ B_{22} \end{bmatrix} = \begin{pmatrix} A_{11} & A_{12} \\ A_{21} & A_{22} \end{pmatrix} \begin{pmatrix} B_{11} & B_{12} \\ B_{21} & B_{22} \end{pmatrix}$$

According to our rules of multiplication of a column vector by a matrix this becomes

$$\mathbb{A} \cdot \mathbb{B} = \begin{bmatrix} A_{11}B_{11} + A_{12}B_{21} \\ \hline A_{21}B_{11} + A_{22}B_{21} \end{bmatrix} \begin{bmatrix} A_{11}B_{12} + A_{12}B_{22} \\ \hline A_{21}B_{12} + A_{22}B_{22} \end{bmatrix}$$

$$= \left(\begin{array}{c|c} A_{11}B_{11} + A_{12}B_{21} & A_{11}B_{12} + A_{12}B_{22} \\ \hline A_{21}B_{11} + A_{22}B_{21} & A_{22}B_{21} + A_{22}B_{22} \end{array}\right)$$

We can dispense with this intermediate viewpoint now and proceed to the general rules governing the multiplication of the two matrices $\mathbb{A}$ and $\mathbb{B}$.

Definition of the product of *Two* Matrices $\mathbb{A}$ and $\mathbb{B}$ ($\mathbb{C} = \mathbb{A} \cdot \mathbb{B}$).

$$\left(\begin{array}{c|c} C_{11} & C_{12} \\ \hline C_{21} & C_{22} \end{array}\right) = \left(\begin{array}{c|c} A_{11} & A_{12} \\ \hline A_{21} & A_{22} \end{array}\right) \left(\begin{array}{c|c} B_{11} & B_{12} \\ \hline B_{21} & B_{22} \end{array}\right) =$$

$$\left(\begin{array}{c|c} A_{11}B_{11} + A_{12}B_{12} & A_{11}B_{12} + A_{12}B_{22} \\ \hline A_{21}B_{11} + A_{22}B_{21} & A_{21}B_{12} + A_{22}B_{22} \end{array}\right)$$

Note that the **elements of the product matrix** are obtained by multiplying a **row** in $\mathbb{A}$ by a **column** in $\mathbb{B}$.

As an example let us compute the product matrix of two particular square arrays.

Let

$$\mathbb{A} = \begin{pmatrix} 1 & 3 \\ -4 & 0 \end{pmatrix}$$

and

$$\mathbb{B} = \begin{pmatrix} 2 & 1 \\ 1 & -2 \end{pmatrix}$$

Then

$$\mathbb{A} \cdot \mathbb{B} = \begin{pmatrix} 1 & 3 \\ -4 & 0 \end{pmatrix} \begin{pmatrix} 2 & 1 \\ 1 & -2 \end{pmatrix} =$$

$$\left(\begin{array}{c|c} 2 + 3 & 1 - 6 \\ \hline -8 + 0 & -4 - 0 \end{array}\right) = \begin{pmatrix} 5 & -5 \\ -8 & -4 \end{pmatrix}$$

It is well to recognize the values assigned to the elements for our special problem above. We have taken a problem in which

$$\begin{array}{llll} A_{11} = 1 & A_{12} = 3 & B_{11} = 2 & B_{22} = 1 \\ A_{21} = -4 & A_{22} = 0 & B_{21} = 1 & B_{22} = -2 \end{array}$$

We can now relate the resulting algebraic relations just obtained to the components of the resultant matrix $\mathbb{C}$.

Note that the *product* of two matrices is itself a *matrix*. Therefore our problem consists of relating the elements A_{lm} and B_{sr} to the elements of the product matrix C_{kj}.

Symbolically we write the product as

$$\mathbb{C} = \mathbb{A} \cdot \mathbb{B}$$

To obtain the elements of $\mathbb{C}$ we multiply **rows** in the **first matrix** by **columns** in the **second matrix.**

The elements of the resulting sums consist of products of elements with like joining **indices.** The outer **indices** serve to designate the position in the final product matrix.

To show this let us reconsider our first example, indicating the multiplications which give the one-one position in the product matrix.

We have called the product matrix

$$\mathbb{C} = \mathbb{A} \cdot \mathbb{B}$$

Thus

$$C_{11} = A_{11} B_{11} + A_{12} B_{21}$$

Notice that each term in the sum has outer indices corresponding to the indices on the $\mathbb{C}$ element. Again notice that the inner indices of each product are alike, and that the number of terms in the sum runs over *all* possible values of the indices. In this case the matrices are 2x2, therefore there are but **two** terms in the sum.

Let us now reconsider our element

$$C_{11} = A_{11} B_{11} + A_{12} B_{21}$$

This sum can be written in a more abbreviated manner as

$$C_{11} = \sum_{k=1}^{k=2} A_{1k} B_{k1}$$

It is perhaps interesting to use the **summation convention** in this particular example

$$C_{11} = \sum_{k=1}^{k=2} A_{1k} B_{k1} = A_{1k} B_{k1}$$

i.e., the existence of the repeated inner index k indicates that this index should be summed over all possible values of k.

There are four components in the product matrix C. We designate any one of the four by the general symbol

$$C_{jm}$$

where j can be 1 or 2 and m can be 1 or 2.

Thus

$$C_{jm} = A_{j1} B_{1m} + A_{j2} B_{2m} = \sum_{k=1}^{k=2} A_{jk} B_{km} = A_{jk} B_{km}$$

As an example we can write out the expansion of C_{21} by taking the case in which $j = 2$ and $m = 1$

$$C_{21} = A_{21} B_{11} + A_{22} B_{21}, \text{ or}$$

$$\begin{pmatrix} A_{11} & A_{12} \\ A_{21} & A_{22} \end{pmatrix} \begin{pmatrix} B_{11} & B_{12} \\ B_{21} & B_{22} \end{pmatrix} = \begin{pmatrix} A_{21} B_{11} + A_{22} B_{21} & \\ & \end{pmatrix}$$

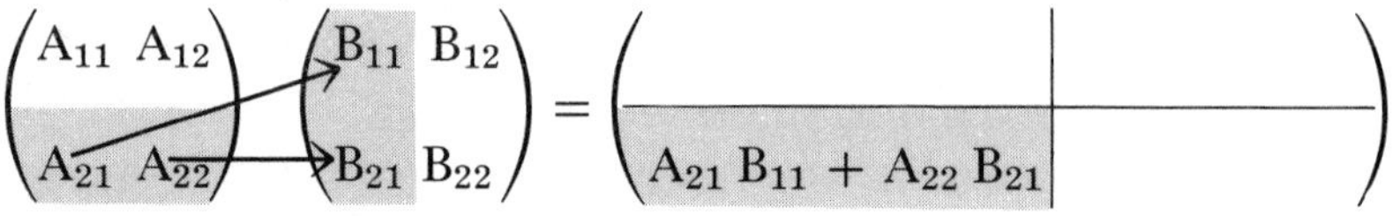

The Inverse Matrix

Up to this point we have considered only that transformation which expresses x_j' as a linear combination of the coordinates x_1 and x_2.

An equally important transformation is *the rotation which carries the coordinates x_1' and x_2' back to x_1 and x_2.*

This is the inverse rotation operation labeled by the symbol $\mathbb{S}^{-1}$ with elements S_{jm}'.

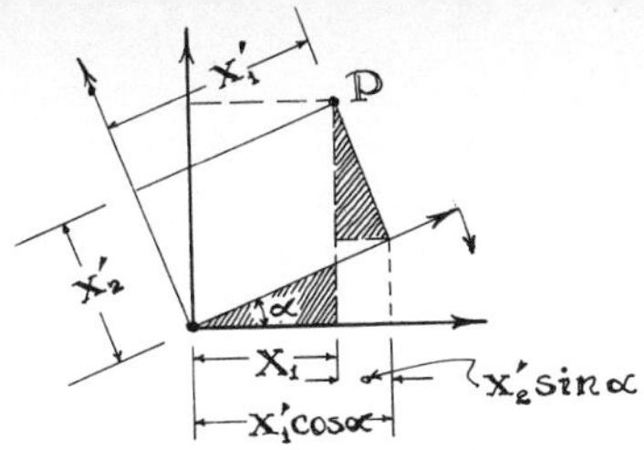

From the first diagram.

$x_1 = (\cos\alpha)\, x_1' + (-\sin\alpha)\, x_2'$

$x_1 = S_{11}' x_1' + S_{12}' x_2'$

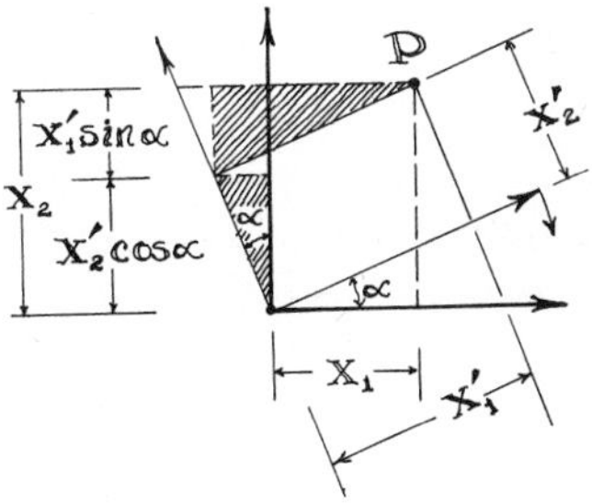

From the second diagram.

$x_2 = (\sin\alpha)\, x_1' + (\cos\alpha)\, x_2'$

$x_2 = S_{21}'\, x_1' + S_{22}'\, x_2'$

Both of these relations could have been obtained by solving for the x's in the two simultaneous algebraic expressions,

$$x_1' = (\cos\alpha)x_1 + (\sin\alpha)x_2$$
$$x_2' = (-\sin\alpha)x_1 + (\cos\alpha)x_2$$

The components S_{jk}' form the elements of the matrix $\mathbb{S}^{-1}$, *the inverse rotation matrix.*

$$S_{11}' = \cos\alpha \quad S_{12}' = -\sin\alpha$$
$$S_{21}' = \sin\alpha \quad S_{22}' = \cos\alpha$$

and

$$\mathbb{S}^{-1} = \begin{pmatrix} \cos\alpha & -\sin\alpha \\ \sin\alpha & \cos\alpha \end{pmatrix}$$

We have defined the product of two matrices $\mathbb{A} \cdot \mathbb{B}$ by showing that an element of the resulting matrix could be written

$$(\mathbb{A} \cdot \mathbb{B})_{lm} = \sum_{\alpha \| k} A_{lk} B_{km} = A_{lk} B_{km}$$

We shall now discover that the product of $\mathbb{S}^{-1}$ and $\mathbb{S}$ has a *unique form.*

$$S^{-1} \cdot S = \begin{pmatrix} \cos\alpha & -\sin\alpha \\ \sin\alpha & \cos\alpha \end{pmatrix} \begin{pmatrix} \cos\alpha & \sin\alpha \\ -\sin\alpha & \cos\alpha \end{pmatrix} =$$

$$\begin{pmatrix} 1 & 0 \\ 0 & 1 \end{pmatrix} = \textit{\textbf{the unit matrix}}$$

The reader should work this out in detail to check the result. Another way of writing the components of $S^{-1} \cdot S$ is

$$(S^{-1} \cdot S)_{jm} = \sum_{k=1}^{2} S_{jk}' \, S_{km} = \delta_{jm}$$

Where δ_{jm} is the *Kronecker delta.*

To remind the reader this symbol has the value ONE when "j" is *equal* to "m", the symbol has the value ZERO when "j" is *not* equal to "m".

$$\delta_{jm} = \begin{matrix} 1 \; (j = m) \\ 0 \; (j \neq m) \end{matrix}$$

The linear algebraic equations specifying the inverse rotation

$$x_1 = (\cos\alpha)\, x_1' + (-\sin\alpha)\, x_2'$$
$$x_2 = (\sin\alpha)\, x_1' + (\cos\alpha)\, x_2'$$

can be written as a matrix product;

$$\begin{bmatrix} x_1 \\ x_2 \end{bmatrix} = \begin{pmatrix} S_{11}' & S_{12}' \\ S_{21}' & S_{22}' \end{pmatrix} \begin{bmatrix} x_1' \\ x_2' \end{bmatrix} = \begin{bmatrix} S_{11}'x_1' + S_{12}'x_2' \\ \hline S_{21}'x_1' + S_{22}'x_2' \end{bmatrix}$$

It is of some interest to see that our first rotation could have been multiplied by S^{-1} to give this result which we obtained geometrically.

$$S^{-1} \cdot \begin{bmatrix} x_1' \\ x_2' \end{bmatrix} = S^{-1} \cdot S \begin{bmatrix} x_1 \\ x_2 \end{bmatrix} = I \cdot \begin{bmatrix} x_1 \\ x_2 \end{bmatrix} = \begin{bmatrix} x_1 \\ x_2 \end{bmatrix}$$

Symbolically this can be written

$$S^{-1} \cdot \mathbf{r}' = S^{-1} \cdot S \cdot \mathbf{r} = I \cdot \mathbf{r} = \mathbf{r}$$

The operation S followed by the operation S^{-1} corresponds to the series of rotations shown below.

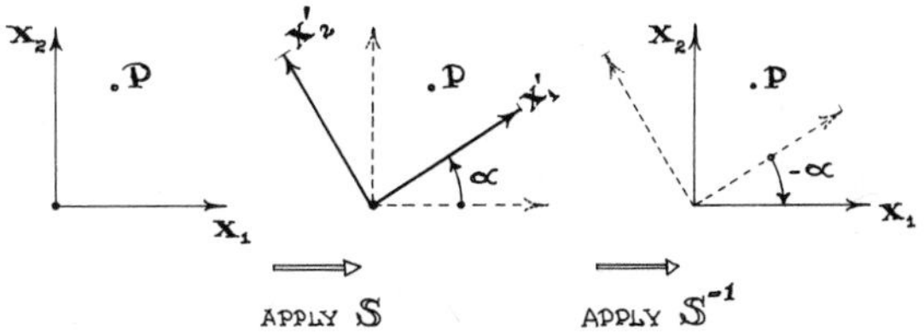

If the magnitude or determinant of the matrix A is *not* zero we define the inverse of A by the product

$$A \cdot A^{-1} = I$$

where I is the unit matrix.

The elements A_{nk}' of A^{-1} can be obtained from the definition of the Determinant. Consider $A \cdot A^{-1} = I$; or

$$\sum_n A_{mn} A_{nk}' = \delta_{mk}$$

The determinant of A is defined by

$$\text{Det } A = \sum_n (-1)^{m+n} A_{mn} \text{ minor } A_{mn}$$

or

$$\delta_{mk} = \sum_n (-1)^{k+n} A_{mn} \frac{\text{min } A_{kn}}{\text{Det } A}$$

From our definition of the inverse

$$\delta_{mk} = \sum_{n} A_{mn} A_{nk}'$$

thus

$$A_{nk}' = \frac{(-1)^{k+n} \text{ minor } A_{kn}}{\text{Det } A}$$

In order to demonstrate a practical application of this elegance, consider as an example of such a calculation the matrix,

$$A = \begin{pmatrix} 3 & 1 & 0 \\ -1 & 2 & 1 \\ 0 & 2 & 2 \end{pmatrix}$$

Now

$$\text{Det } A = \begin{vmatrix} 3 & 1 & 0 \\ -1 & 2 & 1 \\ 0 & 2 & 2 \end{vmatrix} = 3 \cdot \begin{vmatrix} 2 & 1 \\ 2 & 2 \end{vmatrix} - 1 \cdot \begin{vmatrix} -1 & 1 \\ 0 & 2 \end{vmatrix} + 0$$

$$= 3(4 - 2) - 1(-2) = 6 + 2 = 8$$

The reader can check this result by the standard diagonal multiplication of a 3 x 3 determinant. Remember that all determinants of order higher than 3 *must be* expanded by the method of minors.

Notice at this point that the minor of

A_{11} is

$$\begin{vmatrix} 2 & 1 \\ 2 & 2 \end{vmatrix} = +2$$

3	1	0
−1	2	1
0	2	2

The minor of A_{23} is

$$\begin{vmatrix} 3 & 1 \\ 0 & 2 \end{vmatrix} = +6$$

3	1	0
−1	2	1
0	2	2

Thus

$$A_{11}' = (-1)^{1+1}\frac{\min A_{11}}{\text{Det } \mathbb{A}} = +\frac{2}{8} = \frac{1}{4}$$

and

$$A_{32}' = (-1)^{3+2}\frac{\min A_{23}}{\text{Det } \mathbb{A}} = -\frac{6}{8} = -\frac{3}{4}$$

The reader, as an exercise, can now demonstrate that

$$\mathbb{A}^{-1} = \frac{1}{8}\begin{pmatrix} 2 & -2 & 1 \\ 2 & 6 & -3 \\ -2 & -6 & 7 \end{pmatrix}$$

and in addition

$$\mathbb{A} \cdot \mathbb{A}^{-1} = \frac{1}{8}\begin{pmatrix} 3 & 1 & 0 \\ -1 & 2 & 1 \\ 0 & 2 & 2 \end{pmatrix}\begin{pmatrix} 2 & -2 & 1 \\ 2 & 6 & -3 \\ -2 & 6 & 7 \end{pmatrix} = \frac{1}{8}\begin{pmatrix} 8 & 0 & 0 \\ 0 & 8 & 0 \\ 0 & 0 & 8 \end{pmatrix} = \mathbb{I}$$

The Orthogonal Property of $\mathbb{S}$

As indicated previously the interchange of rows and columns of a matrix $\mathbb{A}$, forms the **transposed matrix** $\tilde{\mathbb{A}}$

If

$$\mathbb{A} = \begin{pmatrix} A_{11} & A_{12} \\ A_{21} & A_{22} \end{pmatrix}$$

then

$$\tilde{\mathbb{A}} = \begin{pmatrix} A_{11} & A_{21} \\ A_{12} & A_{22} \end{pmatrix}$$

Another way of writing this is to state that

$$\tilde{A}_{jm} = A_{mj}$$

By examining our rotation matrices $\mathbb{S}$ and $\mathbb{S}^{-1}$ we find that

$$\mathbb{S} = \begin{pmatrix} \cos\alpha & \sin\alpha \\ -\sin\alpha & \cos\alpha \end{pmatrix}$$

and

$$\tilde{\mathbb{S}} = \begin{pmatrix} \cos\alpha & -\sin\alpha \\ \sin\alpha & \cos\alpha \end{pmatrix}$$

with the result that

$$\tilde{\mathbb{S}} = \mathbb{S}^{-1}$$

The transpose of $\mathbb{S}$ is *equal* to the inverse of $\mathbb{S}$. This is a unique property which is the case for a very special class of matrices.

Such a matrix is called **an orthogonal matrix.** An orthogonal transformation is one for which the following relation holds,

$$\tilde{\mathbb{S}}\,\mathbb{S} = \mathbb{I}$$

remembering that

$$\mathbb{S}^{-1} \cdot \mathbb{S} = \mathbb{I}$$

Also note for future reference that the magnitude of an orthogonal matrix is 1.

One important consequence of the orthogonality of the rotation matrix is that it does not change the distance between two points P and Q. A more elegant way of saying this is to state that **the distance between two points is *invariant* under an orthogonal transformation.**

To illustrate this let us consider the distance d from the origin O to the point P (x_1, x_2).

$$d^2 = x_1{}^2 + x_2{}^2$$

We can manufacture d^2 by a matrix multiplication by specifying that when multiplying by a position vector **r** from the left we must use the **transpose** of **r** which will be a row matrix.

$$d^2 = \tilde{\mathbf{r}} \cdot \mathbf{r} = [x_1\ x_2] \begin{bmatrix} x_1 \\ x_2 \end{bmatrix} = x_1^2 + x_2^2 = \textbf{a scalar}$$

To prove the invariance of d^2 under rotation we shall show that

$$d^2 = x_1^2 + x_2^2 = x_1'^2 + x_2'^2$$

Remember that the **inverse rotation** is

$$\begin{bmatrix} x_1 \\ x_2 \end{bmatrix} = \begin{pmatrix} \cos\alpha & -\sin\alpha \\ \sin\alpha & \cos\alpha \end{pmatrix} \begin{bmatrix} x_1' \\ x^{2'} \end{bmatrix}, \text{ or } \mathbf{r} = \mathbb{S}^{-1} \cdot \mathbf{r}'$$

and then the transpose matrices are

$$[x_1, x_2] = [x_1', x_2'] \begin{pmatrix} \cos\alpha & \sin\alpha \\ -\sin\alpha & \cos\alpha \end{pmatrix} \text{ or } \tilde{\mathbf{r}} = \tilde{\mathbf{r}} \cdot \tilde{\mathbb{S}}^{-1} = \tilde{\mathbf{r}} \cdot \mathbb{S}$$

Since[10]

$$d^2 = [x_1, x_2] \begin{bmatrix} x_1 \\ x_2 \end{bmatrix} = \tilde{\mathbf{r}}' \cdot \mathbb{S} \cdot \mathbb{S}^{-1} \cdot \mathbf{r}' = \tilde{\mathbf{r}}' \cdot \mathbb{I} \cdot \mathbf{r}'$$

substitution provides the relation

$$d^2 = [x_1', x_2'] \begin{pmatrix} \cos\alpha & \sin\alpha \\ -\sin\alpha & \cos\alpha \end{pmatrix} \begin{pmatrix} \cos\alpha & -\sin\alpha \\ \sin\alpha & \cos\alpha \end{pmatrix} \begin{bmatrix} x_1' \\ x_2' \end{bmatrix} =$$

$$[x_1', x_2'] \begin{bmatrix} x_1' \\ x_2' \end{bmatrix}$$

and

$$d^2 = x_1'^2 + x_2'^2$$

This completes our proof.

[10] The fact that $\tilde{\mathbb{S}}^{-1} = \mathbb{S}$ is used in these equations. This relation is an equivalent expression of the orthonality.

Special Relativity

In the previous development concerning rotations and matrices we were able to establish a set of formal rules for dealing with linear transformations. As a particular problem the orthogonal two dimensional space rotations were investigated in detail.

The field of relativity in mathematics and physics pertains to space-time transformations. From an abstract point of view the reader with the preceding work as a background might ask:

1) "Are the space-time transformations linear?"

2) "If so, are these transformations orthogonal?"

This is a standard game in the sciences, and this is a game of extrapolation. In other words, if one field of investigation is found to be described by a compact set of mathematical rules; then the first analysis of an adjoining field might well be attempted with the established rules.

Some criticism can be leveled at an attempt to discuss such an extensive field as that of the special relativity. Certainly the laws of transformation of the special relativity can be developed from completely physical arguments, and perhaps one can claim a better understanding of the subject after such a development.

Our purpose is to indicate that the rules for special relativistic transformations can be obtained by the simple assumptions that:

> The speed of light is a constant in all frames of reference moving at constant relative velocities: and that the transformations from one frame of reference to another are **orthogonal.**

In the sections to follow **space-time** transformations will be handled in the same manner as **spatial rotations.** Let us inspect the elementary aspects of the space-time diagram.

Consider an object O (an expensive vase if you wish) of mass M dropped from the corner of a very tall building at time t = O. The object falls

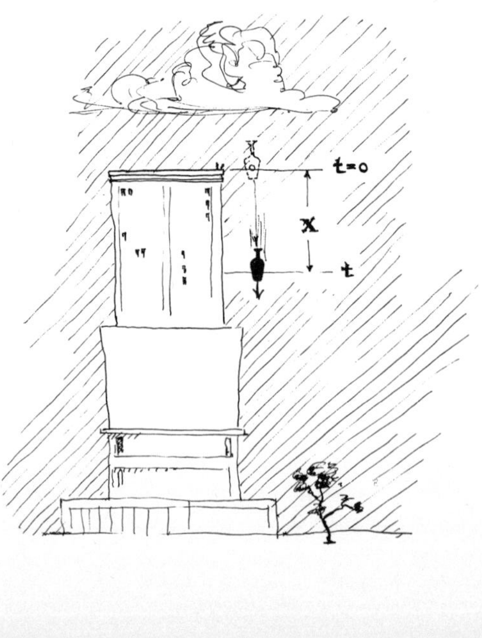

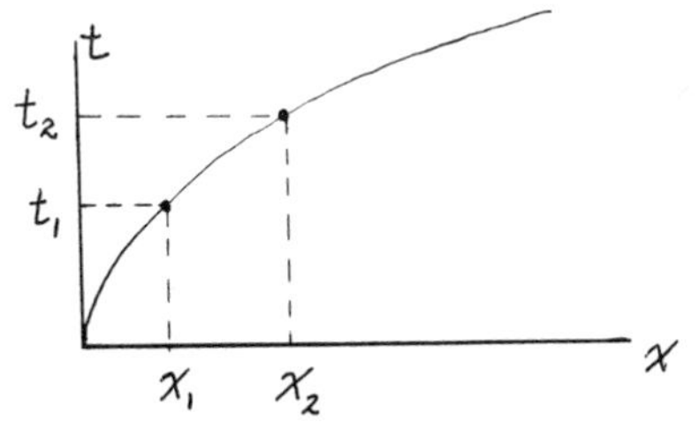

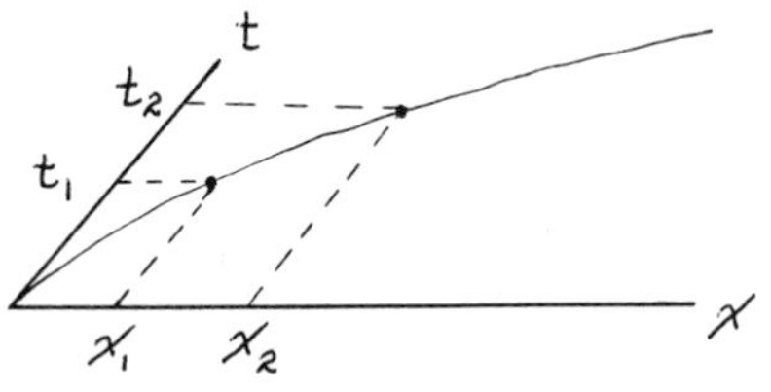

with a constant acceleration g, and at any later time t it has dropped a distance,

$$x = \frac{1}{2} gt^2$$

We can plot the distance of fall x against the time of fall t. The resulting graph is a parabola if the x and t coordinates are drawn mutually perpendicular. There is no reason at this point which demands that x and t be orthogonal. Just as permissible a plot could have been constructed if we had set the x and t axes up with an acute angle between them.

Remember that we used orthogonal axes in configuration space for the convenient reason that the squares of the distance between points could be represented as the sum of the squares of the projections of the intervals between points on the axes.

It will prove convenient to represent the square of the space-time interval between two events as the sum of the squares of the separate space and time intervals.

Notice, we suddenly have introduced the word **"event."** An **event** is a point in the space-time diagram.

In the diagram below a stationary ship at $x_1 = 0$ sends out a light signal at $t = 0$ along the positive x_1 axis. This initiation of a light signal is

"an event" at $x_1 = 0$, and $t = 0$

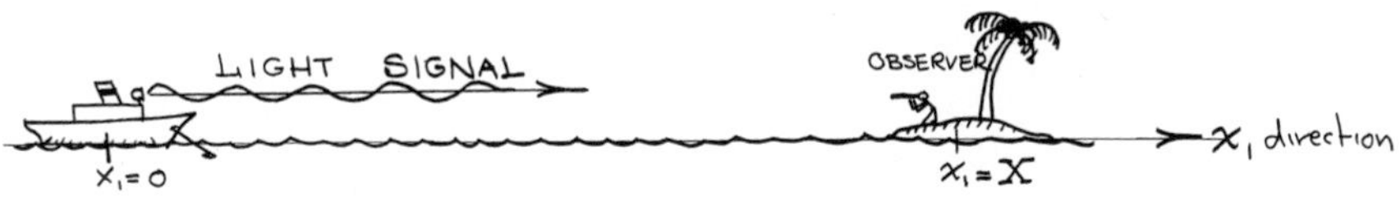

The observer at $x_1 = X$ receives the signal at a time $t = T$. The *reception* of the signal is

"another event" at $x_1 = X$, and $t = T$

If we denote the speed of propagation of light as "c" then

$$c = X/T, \text{ or } X = cT$$

These two events can be plotted upon a space-time diagram.

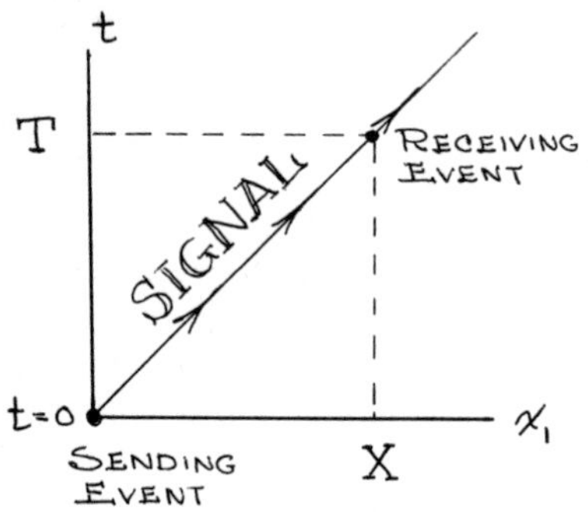

Note that the square of the length of the straight line interval between the two events is

$$X^2 + T^2$$

Since the dimensions of the two quantities are different we can write

$$X^2 + c^2T^2$$

as the square of the length of the interval.

This description of the interval must be further altered.

We proceed by postulating the **Principle of Relativity.**

Consider two coordinate systems O and O′, with O′ moving relative to O with a constant relative velocity V along the x_1 axis.

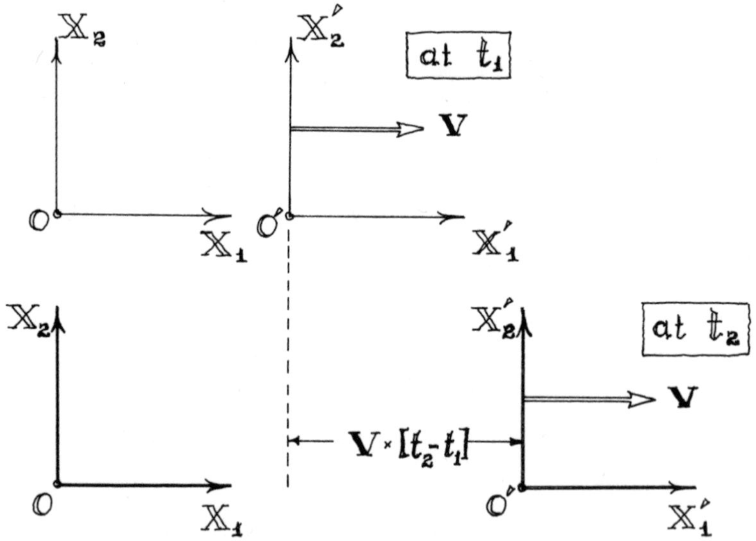

The **Principle of Relativity** states that it is impossible by a physical experiment to label any one coordinate system as intrinsically stationary or absolutely moving with a constant velocity.

One can only detect the presence of relative motion between the two systems.

Consequence!

All physical laws have the same form for systems moving relative to one another with a constant relative velocity. As an example the equations governing the propagation of electromagnetic waves have the same form.

Further Consequence

The speed of propagation of light is the same for all uniformly moving systems, and no effect or signal is propagated with a speed greater than the speed of light.

This principle will seem to lead to a series of physical paradoxes. To explore the results of our postulate let us again consider two frames of reference O and O′ moving relative to one another at a constant velocity V directed along the x_1 axis.

Let us assume that the time-variable in O will be labeled by t and that the time-variable in O′ will be labeled by t′.

At t = O and at t′ = O we can assume that O and O′ coincide and that a light pulse is emitted which diverges from both origins in a spherical wave.

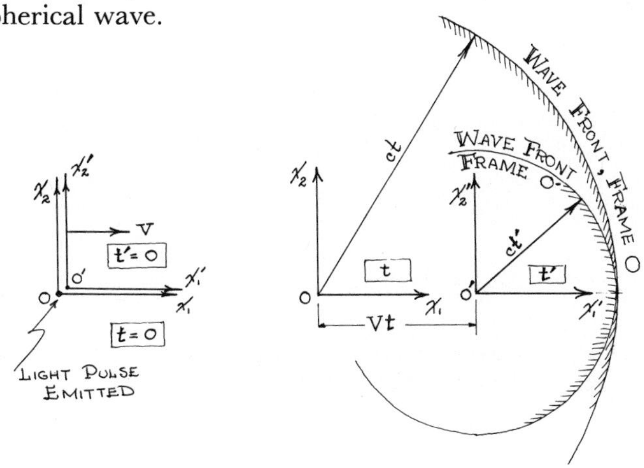

An observer stationary in O sees the light pulse at t as a circle of radius ct.

An observer stationary in O′ sees the light pulse at t′ as a circle of radius ct′.

Because the light pulse is observed as two different circles one of radius ct and the other of radius ct′ we can conclude that the time intervals in O and in O′ are *not* the same.

Let us plot the path of the light signal along x_1 in O and in O′ as a series of **events.**

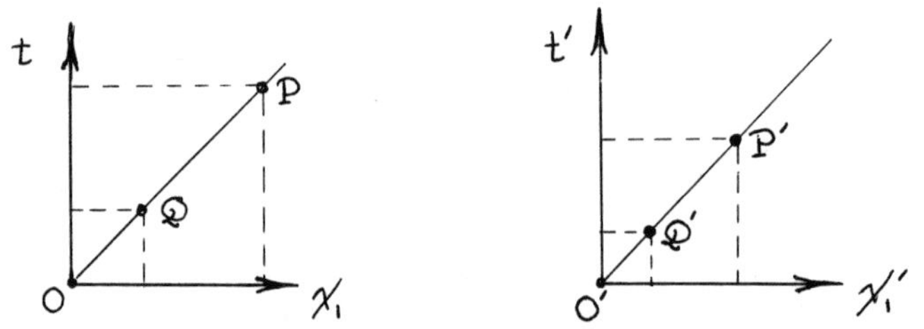

If the light pulse is at x_p at time t_p and at x_q at time t_q then

$$(x_p - x_q) = c(t_p - t_q)$$

i.e.,

$$\text{distance} = \text{speed} \times (\text{time interval});$$

or squaring,

$$\frac{(x_p - x_q)^2}{(t_p - t_q)^2} = c^2$$

P′ and Q′ are the corresponding points in the space of O′. Because the speed of the light pulse is the same in both O and O′ we write

$$c^2 = \frac{(x_p{}' - x_q{}')^2}{(t_p{}' - t_q{}')^2} = \frac{(x_p - x_q)^2}{(t_p - t_q)^2}$$

In our development of spatial rotations we were actually interested in transformations which left the length elements invariant.

We now look for the invariant quantity in the transformation between O and O'. From the preceding equation relating to the square of the velocity of light to the space intervals and time intervals we find that

$$(x_p - x_q)^2 - c^2(t_p - t_q)^2 = (x_p' - x_q')^2 - c^2(t_p' - t_q')^2 = O \quad \textit{or}$$

$$(x_p - x_q)^2 + (ict_p - ict_q)^2 = (x_p' - x_q')^2 + (ict_p' - ict_q')^2 = O$$

where

$$i = \sqrt{-1}$$

If we now redefine the time variable as

$x_2 = ict$ and the space variable x receives the subscript 1, i.e., x_1.

The equation equating events in O and O' becomes

$$(x_{1p} - x_{1q})^2 + (x_{2p} - x_{2q})^2 = (x_{1p}' - x_{1q}')^2 + (x_{2p}' - x_{2q}')^2 = O$$

Now consider any two events M and N the principle of relativity suggests that the distance between the two events expressed in the variables x_1 and x_2 is invariant under a linear transformation. Let the interval between any two events M and N be called $(\tau_m - \tau_n)$, then

$$(\tau_m - \tau_n)^2 = (x_{1m} - x_{1n})^2 + (x_{2m} - x_{2n})^2 = (x_{1m}' - x_{1n}')^2 + (x_{2m}' - x_{2n}')^2$$

By reexamining the problem of the light signal we see immediately that if two events are connected by a light signal, the interval between these events is zero. This result arises from the particular choice of the coordinate x_2 as ict. The important aspect of this result is not the magnitude zero but rather the invariance of the quantity $(\tau_m - \tau_n)^2$ under an orthogonal transformation.

To obtain the **Lorentz transformation** of the **special relativity** it is sufficient to require that the transformation between (x_1, x_2) and (x_1', x_2') be orthogonal.

Imagine two observers O and O′ moving relative to one another with a relative velocity V directed along the x_1 axis. At $t = O$ and $t' = O$ again assume that the origins O and O′ coincide. At a later time t and t′ system O′ has moved to a point $x_1 = Vt$.

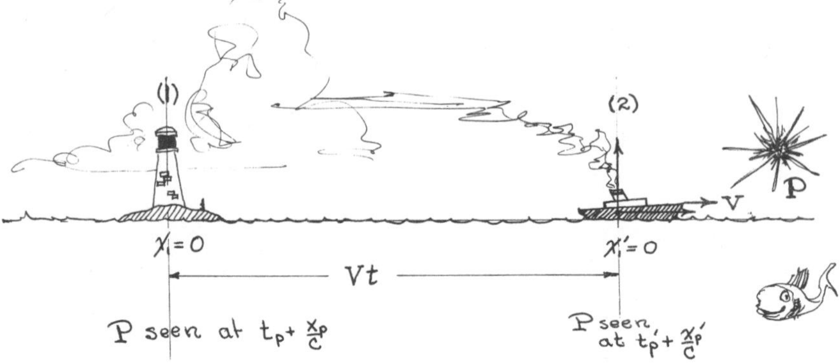

Suppose now that observers O and O′ detect an event (an explosion?) P occurring at x_p and t_p in O and at x_p' and t_p' in O′.

O sees the event at $t_p + x_p/c$
O′ sees the event at $t_p' + x_p'/c$

According to our hypothesis of a linear orthogonal transformation we can relate the two observations by

$$x_1' = S_{11}x_1 + S_{12}x_2$$

and

$$x_2' = S_{21}x_1 + S_{22}x_2$$

To solve these equations for the elements S_{jk} we utilize a special case in which event P occurs at $x_p' = O$ (or directly over the ship);

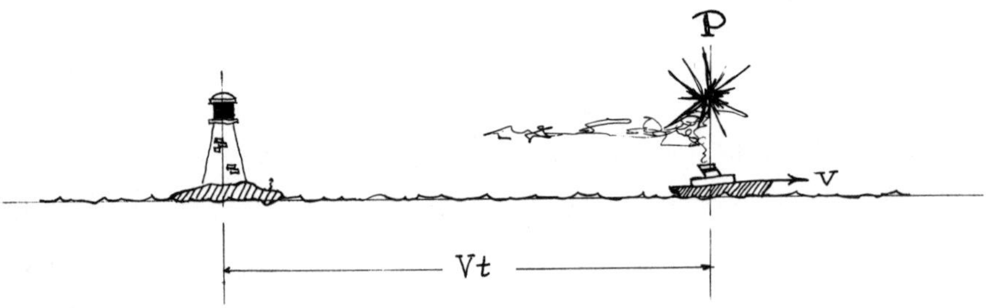

Then: remembering that the position of the ship is at $x_1 = Vt$

$$x_{1p}' = O = S_{11}\,Vt + S_{12}\,(ict)$$

or

$$\frac{S_{12}}{S_{11}} = +\,iV/c$$

Because we assume that the transformation is orthogonal (this is a statement regarding the invariance of the interval $\Delta\tau$ under the transformation),

$$\mathbb{S} \cdot \tilde{\mathbb{S}} = \mathbb{I}$$

or

$$\sum_{k=1}^{2} S_{jk}\tilde{S}_{km} = \delta_{jm}$$

Expanding we find that

$$S_{j1}\tilde{S}_{1m} + S_{j2}S_{2m} = \delta_{jm}$$

Taking specific values of j and m,

$$S_{11}\tilde{S}_{11} + S_{12}\tilde{S}_{21} = S_{11}S_{11} + S_{12}S_{12} = 1$$

or

$$(S_{11})^2 + (S_{12})^2 = 1$$

Solving for S_{11} from $S_{12} = \left(+i\,\frac{V}{c}\right) S_{11}$

$$(S_{11})^2\,(1 - V^2/c^2) = 1$$

or

$$S_{11} = \frac{1}{[1 - V^2/c^2]^{1/2}}$$

and

$$S_{12} = \frac{+\,iV/c}{[1 - V^2/c^2]^{1/2}}$$

Using the orthogonality condition further

$$(j = 1, m = 2, \text{ and } \delta_{12} = 0)$$

$$S_{11}\tilde{S}_{12} + S_{12}\tilde{S}_{22} = S_{11}S_{21} + S_{12}S_{22} = 0$$

Thus

$$\frac{S_{21}}{S_{22}} = \frac{-S_{12}}{S_{11}} = -\,iV/c$$

Finally

$$S_{21}\tilde{S}_{12} + S_{22}\tilde{S}_{22} = 1$$

giving

$$S_{22} = \frac{1}{[1 - V^2/c^2]^{1/2}}$$

and

$$S_{21} = \frac{-\,iV/c}{[1 - V^2/c^2]^{1/2}}$$

When we substitute these values back into the two linear algebraic equations representing the transformation from (x_1, x_2) to (x_1', x_2'), we obtain

$$x_1' = \frac{1}{[1 - V^2/c^2]^{1/2}} \left\{x_1 + i\frac{V}{c}x_2\right\} = \frac{1}{[1 - V^2/c^2]^{1/2}} \{x_1 - Vt\}$$

and

$$x_2' = ict' = \frac{1}{[1 - V^2/c^2]^{1/2}} \left\{-\,i\frac{V}{c}x_1 + x_2\right\}$$

Upon clearing the last of these two equations of the term $i = \sqrt{-1}$ we obtain

$$t' = -\,\frac{(V/c^2)\,x_1 + t}{\sqrt{1 - V^2/c^2}}$$

This is the **Lorentz transformation** of **special relativity.** Our purpose here has not been to discuss the physical basis for these equations as much as it has been to show the manner in which a method developed for the two dimensional Euclidean Space could be applied to a new problem, in this case space-time.

In a manner of speaking we have played an intriguing game. We have asked the question: "Can the matrix method of spatial rotations be applied to space-time transformations"? The answer fortunately is **yes.**

Our method is somewhat different from the customary approach. Ordinarily the Lorentz-Transformation is obtained by a physical argument.

The Lorentz Transformation of the special theory of Einstein was first developed by Lorentz in a manner similar to that used here. Einstein's contribution was to provide a physical description of the equations. We cannot neglect however the insight of Lorentz in first noting the transformation.

Once we have obtained the Lorentz Transformations for x and t, we can predict some of the physical consequences of the special relativity.

The **time intervals** (in two frames moving relative to one another with a constant velocity) are **not the same.** To illustrate this let us consider the two frames O and O′ shown below.

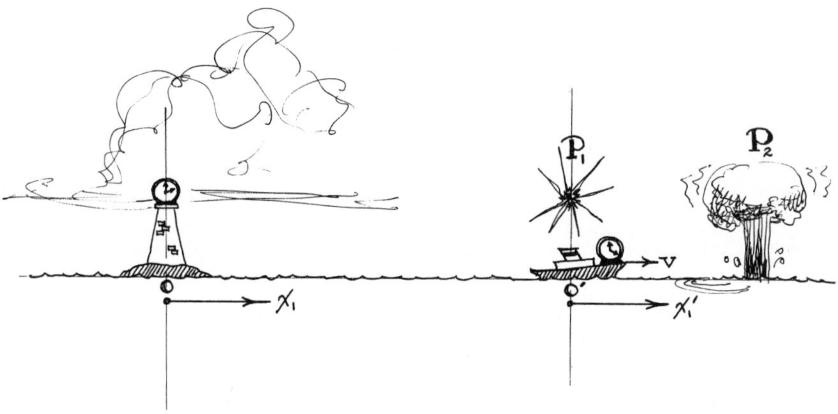

The clock O measures the time t. The clock on O′ measures the time t′. Assuming that both clocks were synchronized to read zero time when the origins O and O′ coincided let us investigate two events P_1 and P_2 observed by both O and O′ (consider that P_2 happens later than P_1 but at the same point in O, i.e., at fixed x_1).

The observer at O records event P_1 at the time t_1 and event P_2 at the time t_2.

The observer on O′ records event P_1 at the time t_1' and event P_2 at the time t_2'.

The observer at O finds that the time interval between P_1 and P_2 is given by

$$\Delta t = \text{change in time} = t_2 - t_1$$

The observer at O′ considers that the time interval is

$$\Delta t' = t_2' - t_1'$$

When we utilize the fact that the point at which P_1 and P_2 occur in O is fixed in O, then using the relation between t′, x_1 and t, we find that

$$\Delta t' = t_2' - t_1' = \frac{t_2 - t_1}{\sqrt{1 - V^2/c^2}} = \frac{\Delta t}{\sqrt{1 - V^2/c^2}}$$

This is called **time dilation,** i.e., Δt′ is greater than Δt.

If the two observers in O and O′ compare time intervals, the observer in O′ concludes that the clock in O is **running slow.** In other words when the observer in O claims one hour between events the observer in O′ finds that more than one hour elapses by his clock. Remember this comparison is conditioned by the fact that the events occur at a fixed point in O. If on the other hand the events occurred at a fixed point in O′ the claims as to slowness of the clocks are reversed.

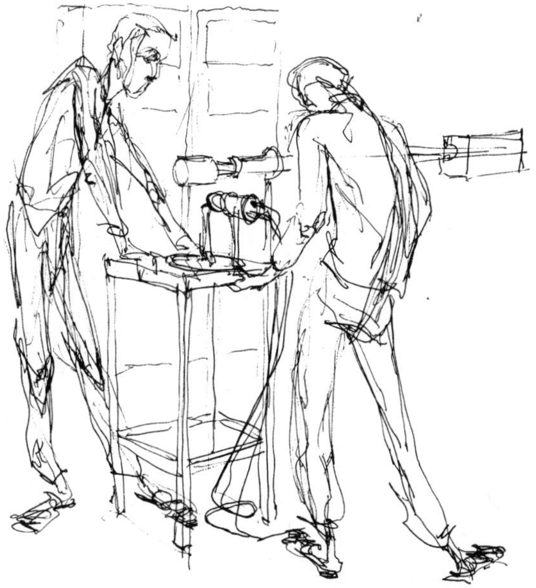

Time dilations are observed in the laboratory. One very spectacular example is that of the **decay of the μ meson.**

The mu minus (symbol μ^-) meson has a mass approximately 200 times the electron mass and has a charge equal to that of the electron. The μ^- meson is unstable in that it decays after a time interval Δt into an electron (e^-) and two neutrinos. The neutrino is a particle of zero mass and is denoted by the symbol ν.

Symbolically the decay is represented as shown below:

$$\mu^- \rightarrow e^- + \nu + \nu$$

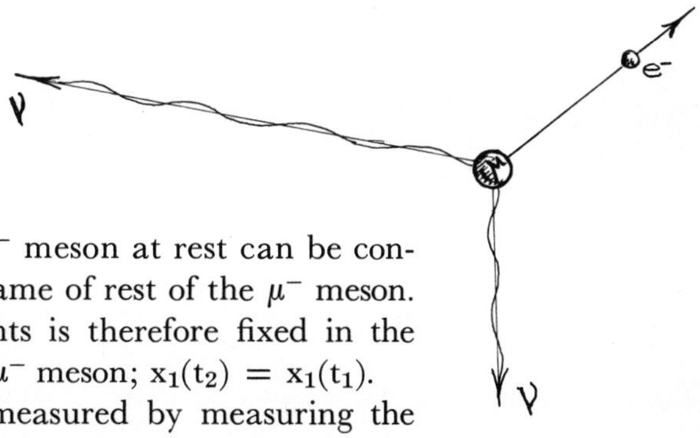

The **lifetime** Δt of the μ^- meson at rest can be considered as fixed in the frame of rest of the μ^- meson. The position of the events is therefore fixed in the frame of reference of the μ^- meson; $x_1(t_2) = x_1(t_1)$.

The **half-life** can be measured by measuring the number of electrons emitted from a beam of mu mesons in a specified time interval, $\Delta t = t_2 - t_1$.

Time dilation is exhibited by the observed fact that the half-life of the mu meson in flight (i.e. moving) is *longer* than the half-life of the stationary mu meson (i.e. at rest in the laboratory frame).

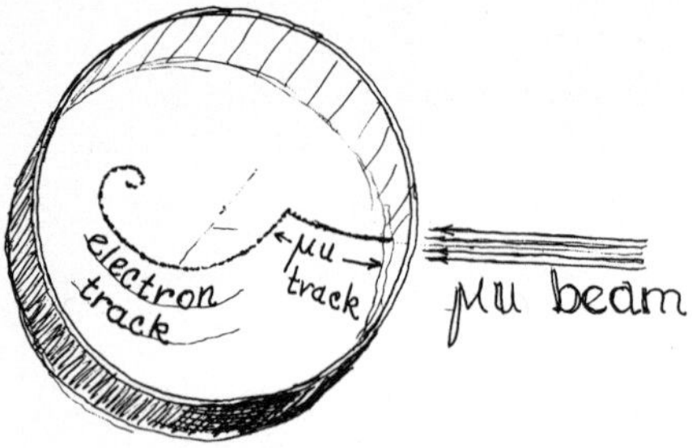

CLOUD CHAMBER

Consider a segment of a μ^- meson beam as the frame O. In a time Δt, N electrons are emitted. In the laboratory frame O′ (moving backward) these N electrons are counted in a time $\Delta t'$ which is greater than or equal to Δt depending upon the relative velocity V.

Experimentally one finds that the stationary (relative velocity zero) mu meson has a shorter half-life than the moving mu meson. If the relative velocity of the mu—meson with respect to the laboratory is V then the life-time Δt of the stationary mu is $\{1 - V^2/c^2\}^{1/2}$ times the lifetime $\Delta t'$ of the moving mu as observed in the laboratory.

$$\Delta t' = \frac{\Delta t}{\sqrt{1 - V^2/c^2}}$$

CHAPTER 2

Vector Algebra

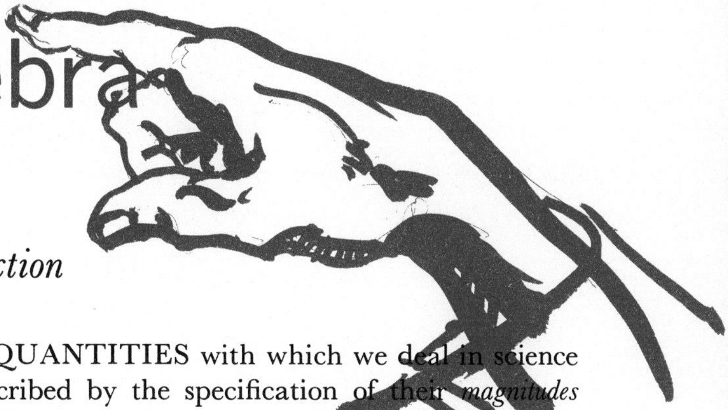

A. Introduction

MANY OF THE QUANTITIES with which we deal in science are adequately described by the specification of their *magnitudes* only. The magnitude of a quantity is a number and as such is classified as a scalar quantity. For example such physical quantities as mass, volume, temperature are scalar quantities.

Often physical quantities require the specification of three numbers to describe them adequately. These three numbers as we will see shortly can refer to a directed line segment (in configuration space). When the distance d between two points P and Q in the space of E_3 was discussed, care was exercised to treat only the square of d. This procedure obviously enabled us to treat the scalar distance between the two points without reference to a preferred direction of the line segment.

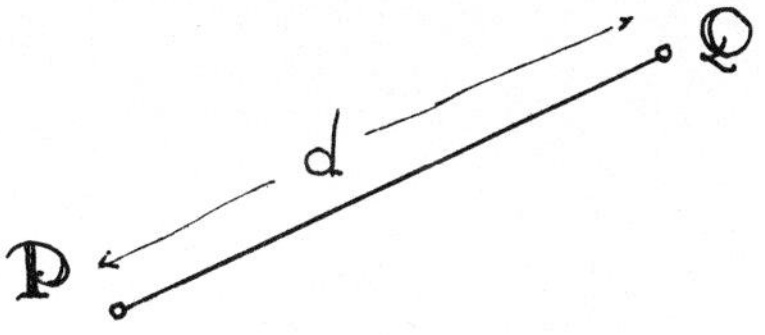

Quantities such as position relative to a specified origin; velocity; acceleration; and force are all directed quantities and as such are ***vectors.***

B. Definition of the Vector

Let P be a fixed point in the Euclidean space E_3 (or E_2). Then by a vector **A** *at* P we mean simply a straight line segment drawn from P to some other point Q with the direction indicated toward Q. We sometimes indicate this vector by $\overrightarrow{PQ}$ instead of **A.**

If Q lies upon P, or P = Q; then the vector is a **null vector** or **zero vector.**

Since the same line segment could equally form a vector at Q pointing in the direction of P, it is necessary to specify the direction of the line segment. In drawing diagrams we usually attach an arrow head to specify the direction. For example as before the vector **A** = $\overrightarrow{PQ}$ *at* P is drawn as

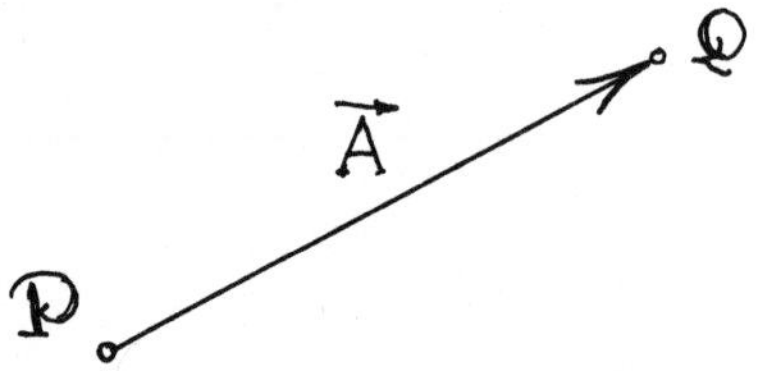

The distance from P to Q is called the *length* or *magnitude* of the vector **A** and is written

$$|\mathbf{A}| \text{ or } |\overrightarrow{PQ}|$$

The *two vertical lines* bracketing a vector indicate that we are considering the magnitude of the vector. Because the magnitude of the vector $\overrightarrow{PQ}$ is equal to the distance between P and Q the magnitude is a SCALAR quantity. A scalar quantity is a more general class of objects being either positive or negative. A MAGNITUDE on the other hand has the limitation that it is always taken *positive.*

C. Properties of Vectors

In this section some of the general properties of vectors will be specified.

Two parallel line segments of equal lengths and having the same direction are said to be ***equal vectors.***

$\mathbf{A} = \mathbf{B}$ in the diagram.

Two parallel line segments of equal lengths but pointing in opposite directions have the property that one is the ***negative*** of the other

$$\mathbf{A} = -\mathbf{C}$$

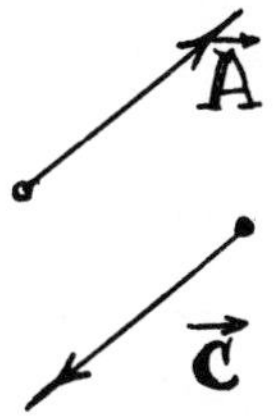

Let **A** be a vector at the point P. In general a vector can be multiplied by a number n (a scalar). This operation produces a vector which has the same direction (or opposite if n is a negative number) as the original vector **A,** but has a magnitude n times as great as the magnitude of the original vector.

The product of the vector **A** with the scalar n is represented by

$$n\mathbf{A}$$

n**A** is parallel to **A** if $n > 0$

n**A** is antiparallel to **A** if $n < 0$

As an illustration consider **A,** 2**A,** and $-\frac{3}{2}\mathbf{A}$

From the definition

$$|n\mathbf{A}| = |n|\ |\mathbf{A}|$$

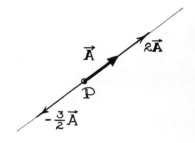

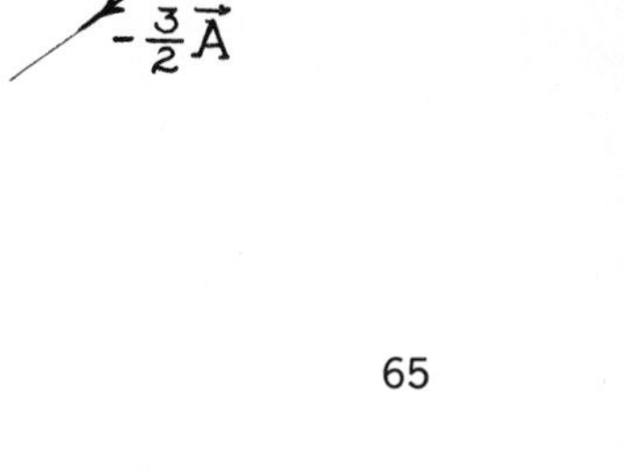

stipulating that the magnitude of a vector is always taken as a positive number. The magnitude sign about the scalar "n" removes any negative signs.

If m is any other number, we can form the product of m and nA in any combination that we desire

$$m\,(n\mathbf{A}) = (mn)\,\mathbf{A} = (mn\mathbf{A})$$

The Unit Vector

If in forming n**A** we take n to be

$$n = \frac{1}{|\mathbf{A}|}$$

assuming that A ≠ O; then the length of the resulting vector after multiplication is *one*.

$$\frac{1}{|\mathbf{A}|}\mathbf{A} = \frac{\mathbf{A}}{|\mathbf{A}|}$$

This vector has the same direction as **A** however

$$\left|\frac{\mathbf{A}}{|\mathbf{A}|}\right| = \frac{|\mathbf{A}|}{|\mathbf{A}|} = 1$$

The unit vector is dimensionless.

Addition of Vectors

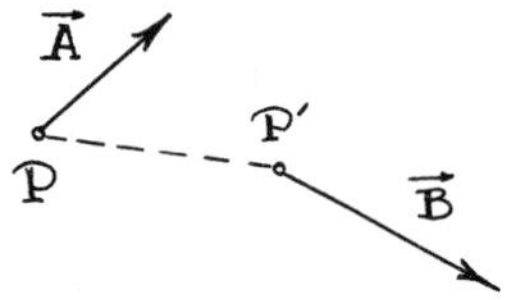

The operations to follow will involve *two* or more vectors; instead of a vector and a number. In these operations involving more than one vector **the vectors must all be located at the same point of** E_3 (or E_2).

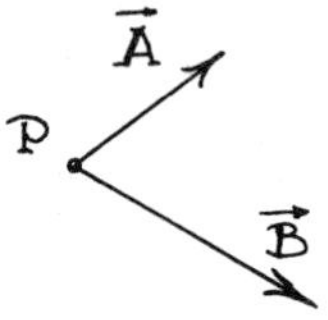

For example it does not make sense to add a vector at P to a vector at a different point P′.

Let **A** and **B** be two vectors located at a

point P. We are going to define a THIRD vector at P called the sum of **A** and **B**, and indicated by **A** + **B**.

To do this construct the *parallelogram* having **A** and **B** as its two sides and let Q be the vertex of the figure opposite from P. Then $\overrightarrow{PQ}$ is by definition the vector **A** + **B**. This is illustrated in the diagram below.

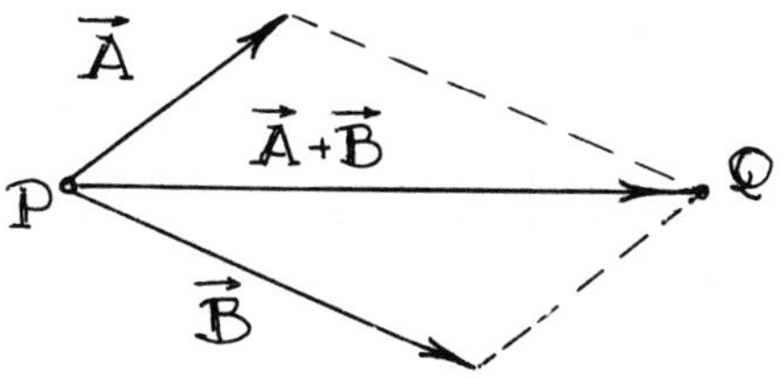

Of course if **A** and **B** happen to lie in the same or opposite directions, then it is not possible to construct the parallelogram. In such a case **A** + **B** is defined in the most obvious way: If **A** and **B** are parallel, then **A** + **B** is defined by the vector of length |**A**| + |**B**| in the common direction of **A** and **B**. Note in the diagram below how this definition is consistent with the parallelogram method as the parallelogram collapses into a line.

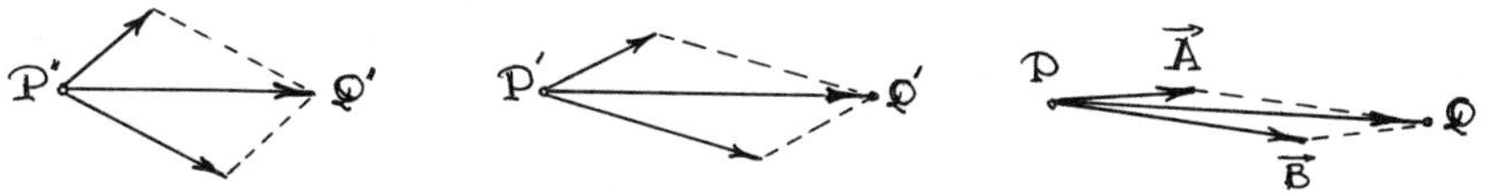

If **A** and **B** are antiparallel, and if say |**A**| > |**B**| then **A** + **B** is the vector of length |**A**| − |**B**| in the direction of **A**. Again this definition is consistent with the parallelogram method.

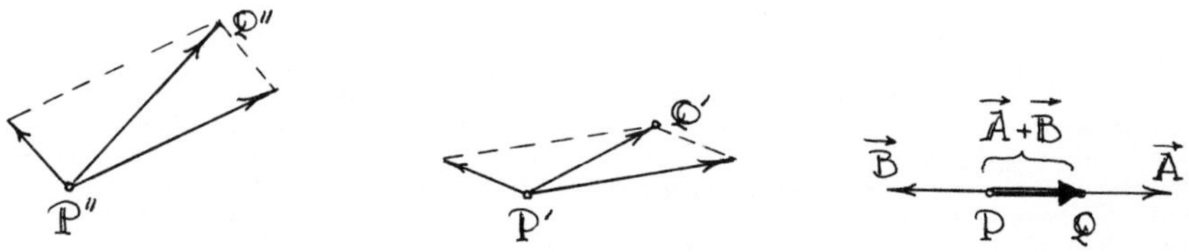

1. Vector Addition obeys the **commutative law**, i.e., the order in which the vectors are added does not alter the final result.

$$\mathbf{A} + \mathbf{B} = \mathbf{B} + \mathbf{A}$$

2. Also Vector Addition obeys the **associative law,** i.e., the grouping in the addition of three or more vectors does not alter the final result:

$$\mathbf{R} = (\mathbf{A} + \mathbf{B}) + \mathbf{C} = (\mathbf{C} + \mathbf{A}) + \mathbf{B} = \mathbf{A} + (\mathbf{B} + \mathbf{C})$$

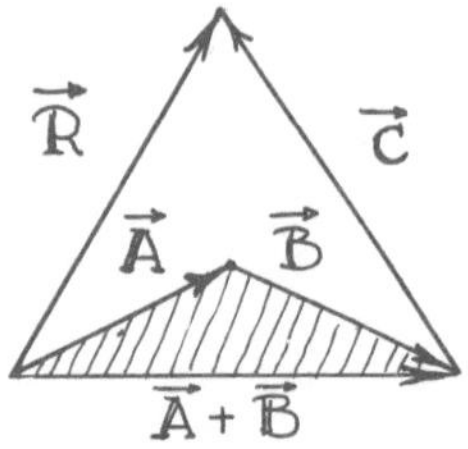

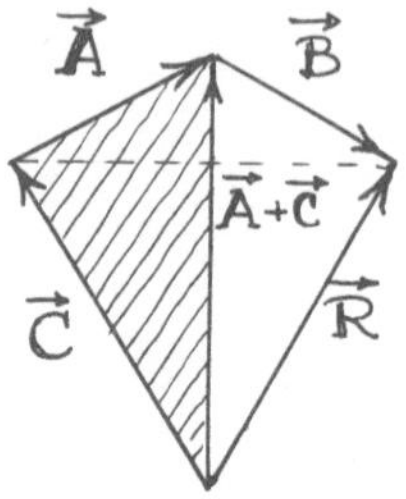

3. Vector Addition obeys the **distributive law,** i.e., a scalar times the sum of two vectors is equal to the sum of the scalar times each vector. Let n be a scalar, then

$$n(\mathbf{A} + \mathbf{B}) = n\mathbf{A} + n\mathbf{B}$$

If A is different from zero and,

$$n\mathbf{A} = O$$

Then n must be zero.

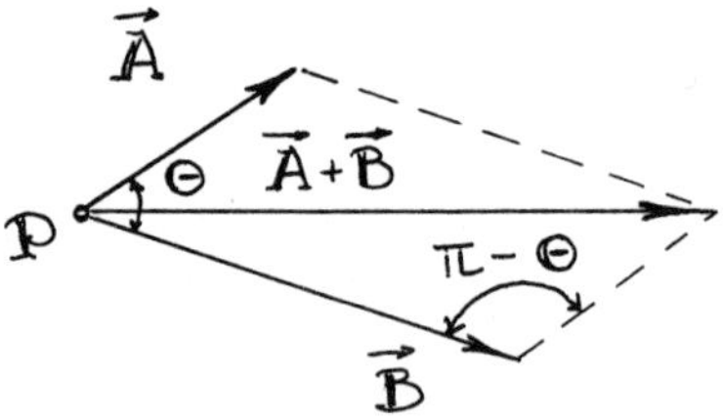

The length of resultant **A** + **B** can be obtained from the law of cosines.

$$|\mathbf{A} + \mathbf{B}| = \{\mathbf{A}^2 + \mathbf{B}^2 - 2\,|\mathbf{A}||\mathbf{B}|\cos(\pi - \theta)\}^{1/2}$$

where θ is the angle between **A** and **B.**

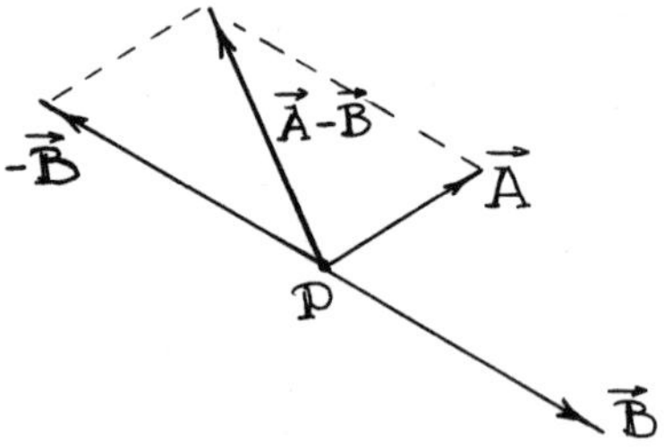

The difference of two vectors is obtained by the rules of addition. Consider the vectors **A** and **B,** and compute **A** − **B.** The difference can be written **A** + (−1) **B.** The multiplication of **B** by (−1) produces a new vector antiparallel to **B** but of equal magnitude.

The relative velocity problem can be utilized as an example of the addition of vectors. The velocity of a moving body is a vector at any instant of time because the velocity has both direction and magnitude (speed).

Velocity is always defined relative to a given coordinate system. For instance we can stipulate that a ship moves southeast at say 20 miles per hour. The coordinates are the spherical grid laid out on the earth's surface.

For convenience let us take the direction south as the x axis[1] and the direction east as the y axis.[1] Then $\mathbf{V}_s$ can be represented graphically as shown to the right.

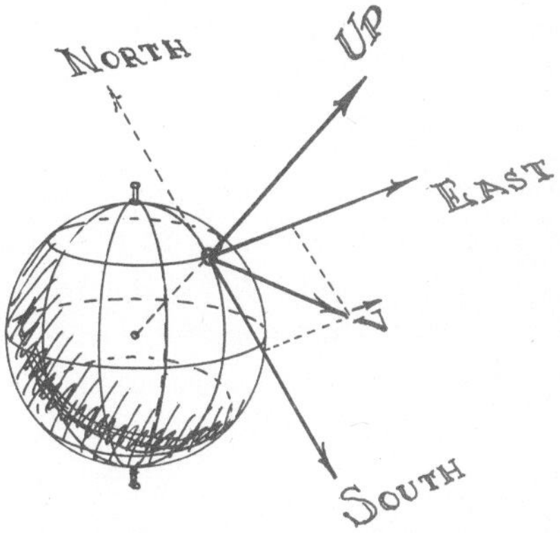

Since velocity implies the motion of a specified object (a ship in this case) relative to a specified frame (the earth's surface in our example) we can label the vector representing the velocity $\mathbf{V}_s$ in such a manner that this information is seen graphically. We place the name of the moving body at the head of the vector, and we label the tail of the vector with the frame of reference.

In our problem involving the ship we label the vector as shown.

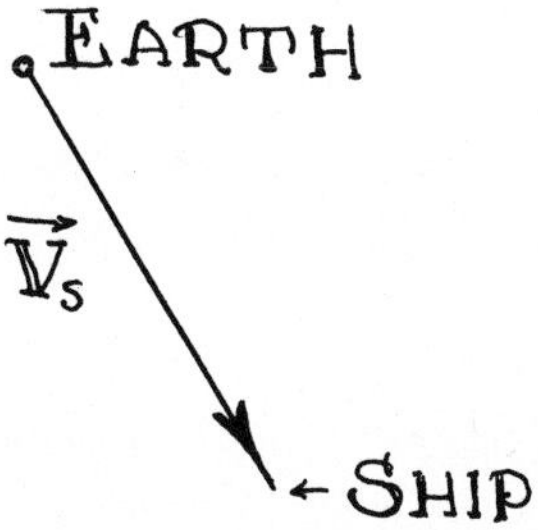

Now suppose we are given the velocity of the wind relative to the ship and we state that the wind blows to the Northeast at a relative velocity of 10 miles per hour (denote this velocity as $\mathbf{v}_w$).

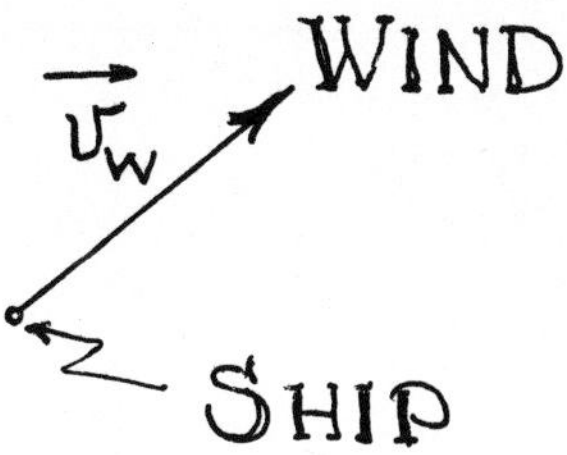

[1] This convention is chosen to correspond to the system of base vectors in spherical coordinates.

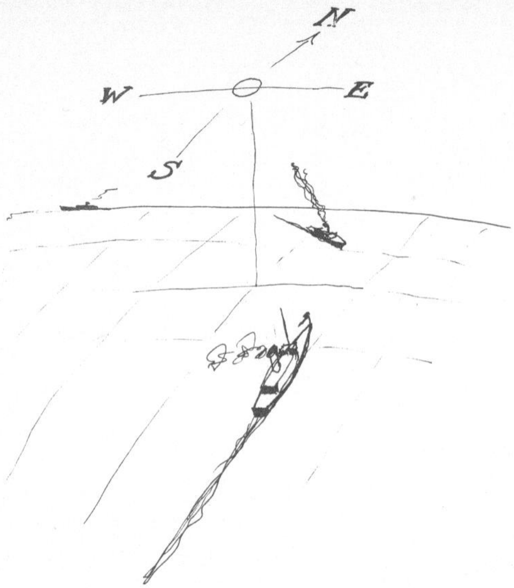

In this case the ship is the frame of reference. Thus the vector giving the relative wind velocity is drawn below.

Up to this point we have defined symbols. What problem can be solved utilizing these symbols? One problem is the determination of the velocity of the wind relative to the earth.

The solution is simple when we add the vectors in such a manner that the labels coincide.

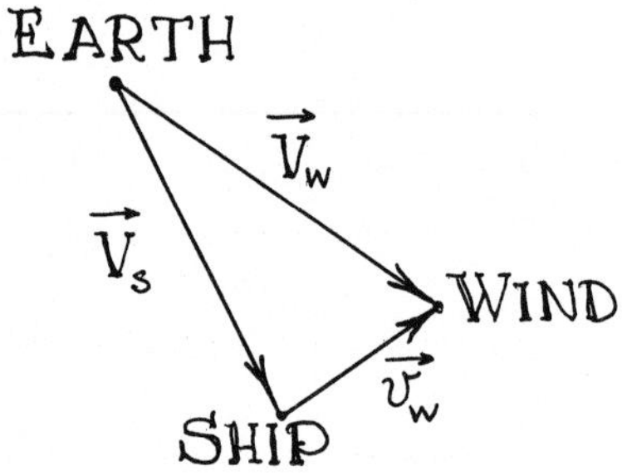

We see that the velocity of the wind relative to the earth $\mathbf{V_W}$ is given by the equation,

$$\mathbf{V_s} + \mathbf{v_W} = \mathbf{V_W}$$

or as it is usually presented

$$\mathbf{v_W} = \mathbf{V_W} - \mathbf{V_s}$$

This approach can now be extended to a more complicated problem. Consider our ship in the problem as ship 1 and now assume that we have a second ship moving due East at a speed of 25 miles per hour.

We can ask the question: "If the wind velocity relative to ship 1 is given ($\mathbf{v_{W_1}}$), what is the velocity of the wind relative to ship 2, $\mathbf{v_{W_2}}$"? The solution is obtained from the vector diagram.

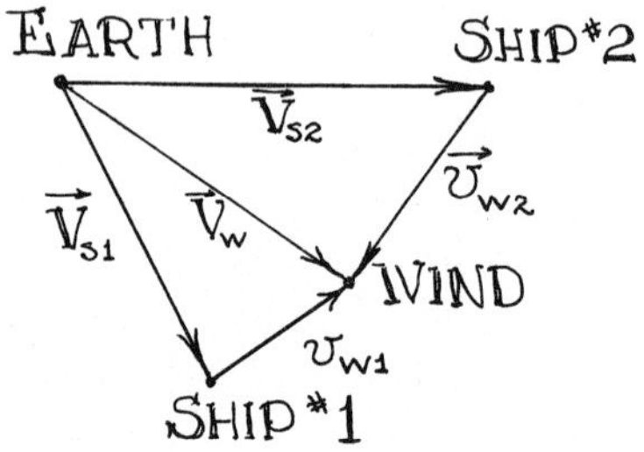

D. *Multiplication of Vectors*

The projection of one vector upon another

Let **A** and **B** be two vectors at P. We assume that $\mathbf{A} \neq O$.

The **projection** of **B** upon **A** is *defined* to be the vector a**A,** where $a = \frac{|\mathbf{B}|}{|\mathbf{A}|} \cos\theta$; θ being the angle between **A** and **B.**

The projection a**A** has the same direction as **A** if $a > 0$ (i.e., if $\cos\theta \geq 0$, and $\theta \leq 90°$). The projection has the opposite direction if $a < 0$ (i.e., $\theta < \pi$ and $\theta > 90°$). The two cases are illustrated below.

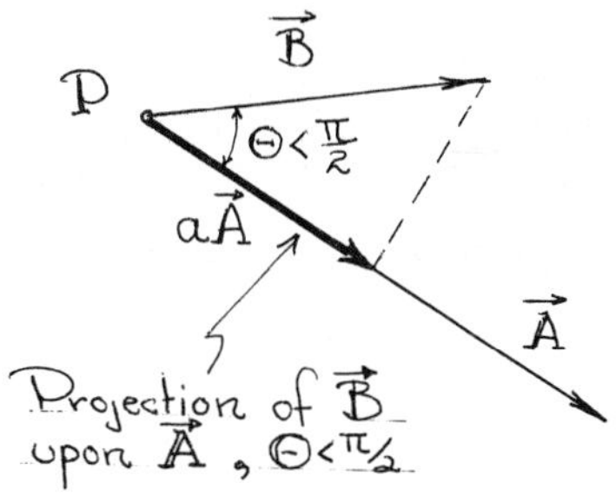

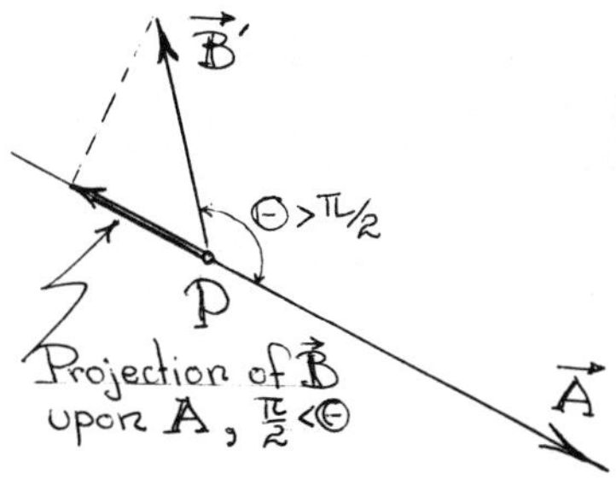

The Scalar or Inner Product

The Scalar Product of two Vectors **A** and **B** can be defined in terms of the ***magnitude*** of the projection of one vector on the other.

The **scalar product** between **A** and **B** is denoted by placing a ***dot*** between the two and is defined as

$$\mathbf{A} \cdot \mathbf{B} = |A|\ |B| \cos\theta$$

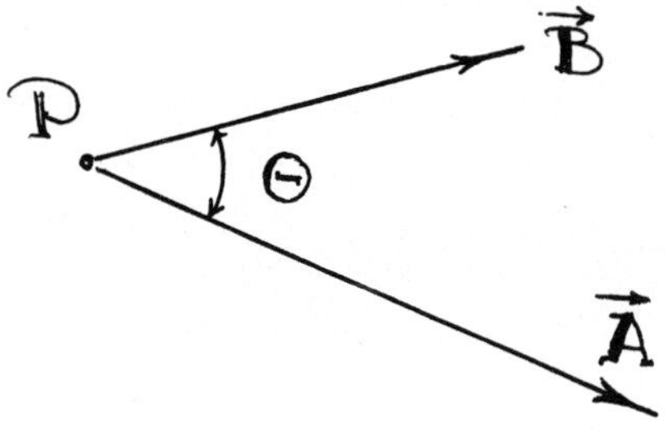

θ is the smallest angle between **A** and **B.**

The **dot** or **scalar product** gives a scalar (not a vector).

This multiplication is convenient to define the magnitude of a vector or a sum of vectors. We find that

$$|\mathbf{A}| = \sqrt{\mathbf{A} \cdot \mathbf{A}}$$

the positive square root of the scalar product of a vector with itself is the magnitude of the vector.

Using this relation we can show that

$$|\mathbf{A} + \mathbf{B}| = \sqrt{(\mathbf{A} + \mathbf{B}) \cdot (\mathbf{A} + \mathbf{B})} = \sqrt{|\mathbf{A}|^2 + |\mathbf{B}|^2 + 2\mathbf{A} \cdot \mathbf{B}}$$

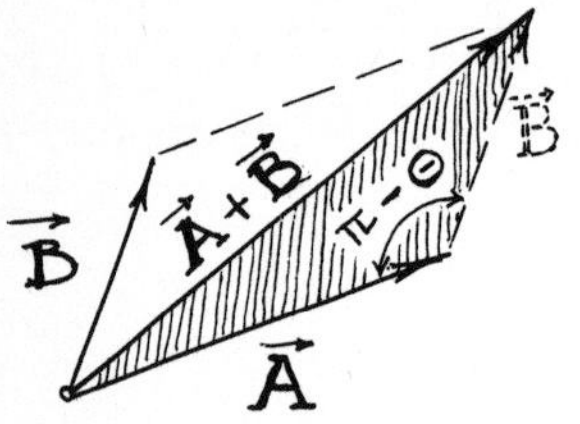

The expression above is a restatement of the law of cosines.

$$|\mathbf{A} + \mathbf{B}|^2 = |\mathbf{A}|^2 + |\mathbf{B}|^2 + 2\,|\mathbf{A}|\,|\mathbf{B}| \cos\theta$$

The order of multiplication of the vectors in the scalar product does not affect the result. The scalar product of *two vectors* commutes.

$$\mathbf{A} \cdot \mathbf{B} = \mathbf{B} \cdot \mathbf{A}$$

If "a" is any number, then

$$(\mathrm{a}\mathbf{A}) \cdot \mathbf{B} = \mathbf{A} \cdot (\mathrm{a}\mathbf{B}) = \mathrm{a}(\mathbf{A} \cdot \mathbf{B})$$

If **C** is another vector at P, then

$$\mathbf{A} \cdot (\mathbf{B} + \mathbf{C}) = \mathbf{A} \cdot \mathbf{B} + \mathbf{A} \cdot \mathbf{C}$$

It follows easily from the definition that the projection of **B** on **A** is (see the previous section)

$$\mathrm{a}\mathbf{A} = \frac{(\mathbf{A} \cdot \mathbf{B})}{|\mathbf{A}|^2}\,\mathbf{A}$$

The work done by a force **F** acting through a distance is defined in terms of the scalar product between the vector **F** and the straight line distance designated by the vector **l**

$$\mathbf{W}_l = \mathbf{F} \cdot \mathbf{l}$$

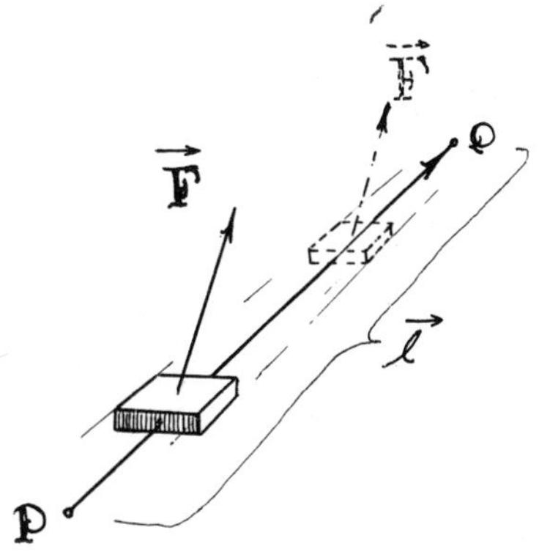

If we consider a body moving between points P and Q in a straight line and subject to a constant force **F**,

then

$$W_1 = \mathbf{F} \cdot \mathbf{l}$$

In practice we need to know the total work done on a body moving along a curve (not in general a straight line) and subject to a force which varies in direction and magnitude from point to point on the curve. A typical case is illustrated below. In this illustration we break the curve into a finite number of segments and draw the force at each point of division.

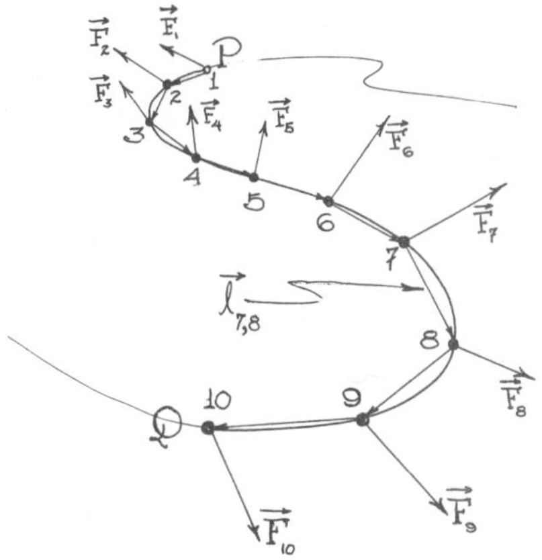

We can obtain an approximate value for the work by taking the sum of the scalar products of the force at the beginning of an interval with a vector cord which approximates the interval.

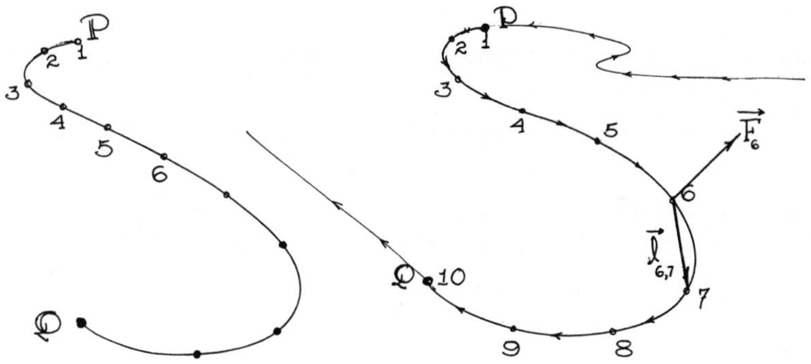

Consider the $6 \rightarrow 7$ interval

$$W_{6\rightarrow 7} \cong \mathbf{F}_6 \cdot \mathbf{l}_{6,7}$$

The *approximation* of the **total work** is obtained by **summing over all of the intervals.**

$$W_{\text{Total}} \simeq \sum_{j=1}^{9} \mathbf{F}_j \cdot \mathbf{l}_{j,j-1}$$

The approximation becomes better as the intervals are taken smaller in length and as the total number of intervals is increased. The proof of this statement is contained in the fact that the cord of a small segment approximates its corresponding arc to a higher degree than does the cord of a larger segment.

Later in the discussion of the integral calculus we shall show that the total work is obtained exactly if we let the length of the intervals approach zero while the number of intervals increases in an appropriate manner.

$$W_{\text{Total}} = \underset{\substack{l\rightarrow 0\\ N\rightarrow\infty}}{\text{limit}} \sum_{j=1}^{N} \mathbf{F}_j \cdot \mathbf{l}_{j,j+1} = \int_P^Q \mathbf{F} \cdot d\mathbf{l}$$

The Cross Product or Vector Product

All of the vector operations which have been discussed up to this point can be performed in either E_2 or E_3. The operation which carries the designation **"vector product"** applies only to vectors in E_3.

We shall associate with any two vectors **A** and **B** (at a point P of E_3) a **third vector A × B** at P. This vector, **A × B,** is called the **cross** (or Vector) **product** of **A** and **B.** This operation is somewhat more complicated than the operations defined previously in that the concept of a right handed system must be invoked.

A vector **A** and a vector **B** form a **plane.**

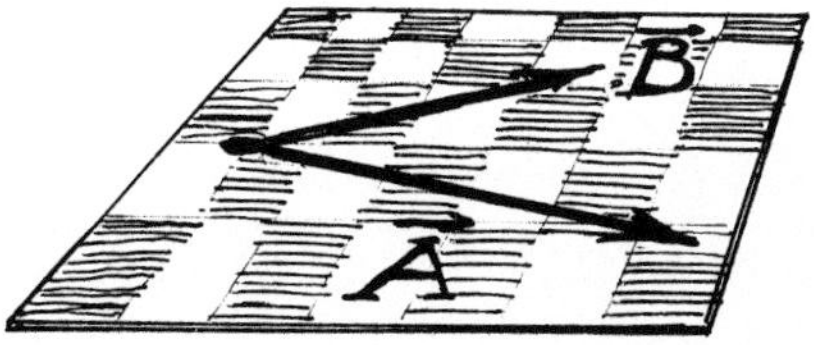

The **angle** θ between **A** and **B** is **measured** in the **plane** of **A** and **B.** Suppose for the moment that $\theta \neq 180°$ or $0°$.

The **perpendicular** to the **plane** of **A** and **B** is a **line characteristic of the combination A** and **B.**

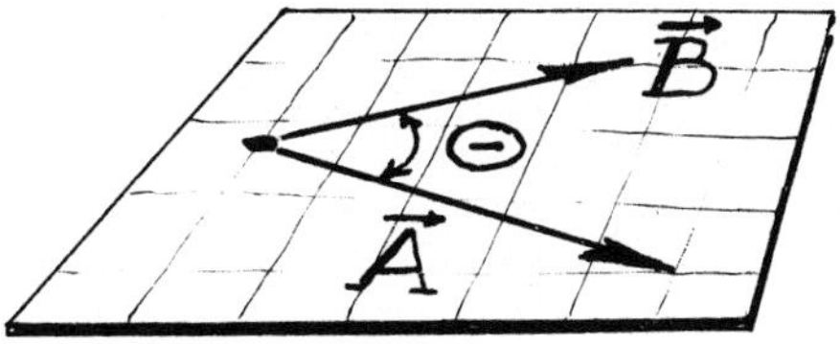

We define the cross product in terms of the properties of the combination **A** and **B** which are listed above.

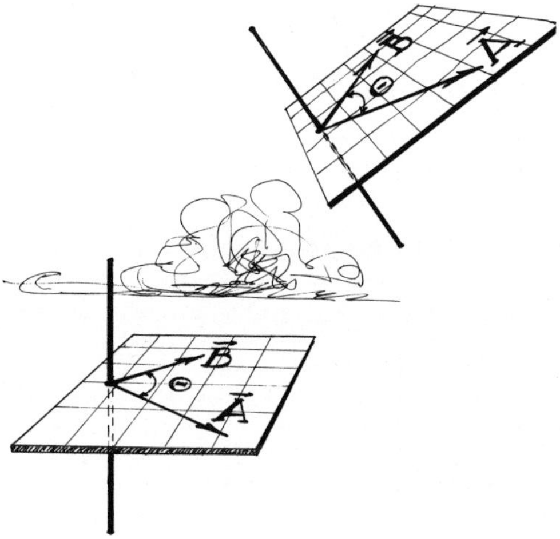

The magnitude of the vector product of **A** and **B** is defined as the magnitude of **A** times the magnitude of **B** times the sine of the (smallest) angle between them,

$$|\mathbf{A} \times \mathbf{B}| = |\mathbf{A}|\ |\mathbf{B}| \sin\theta$$

The **direction** of the **vector product** is taken **perpendicular** to the **plane** defined by **A** and **B.** We have two choices for the direction and the choices are of opposite sense.

The direction is **defined** by a **right hand convention.** Start with the fingers of the right hand pointing in the direction of the ***first*** vector **A.** The fingers are then rotated TOWARD the ***second*** vector **B** by closing the fist. **The thumb then indicates** the **direction** of the **cross product.**

$$\mathbf{C} = \mathbf{A} \times \mathbf{B}$$

$$|\mathbf{C}| = |\mathbf{A}|\ |\mathbf{B}| \sin\theta$$

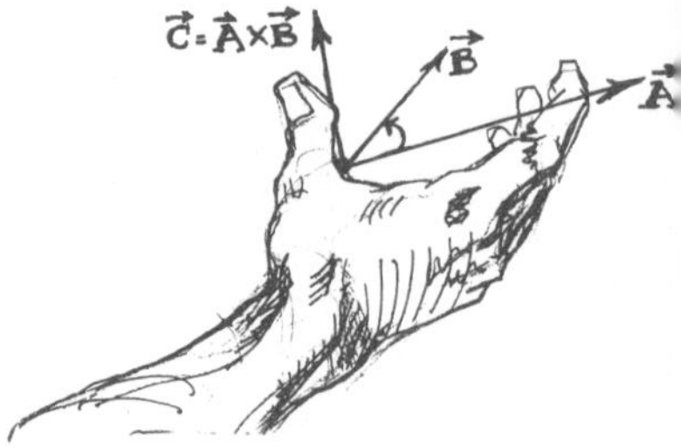

If we utilize our previous definition of a **right handed coordinate system** the direction of the vector **A** × **B** is determined. We align the first vector (**A** in the cross product) with the x axis. The plane of **A** and **B** is oriented in the xy plane. In such a case the vector **A** × **B** is determined by measuring a distance |**A**| |**B**| sin θ **along the positive** Z **axis.**

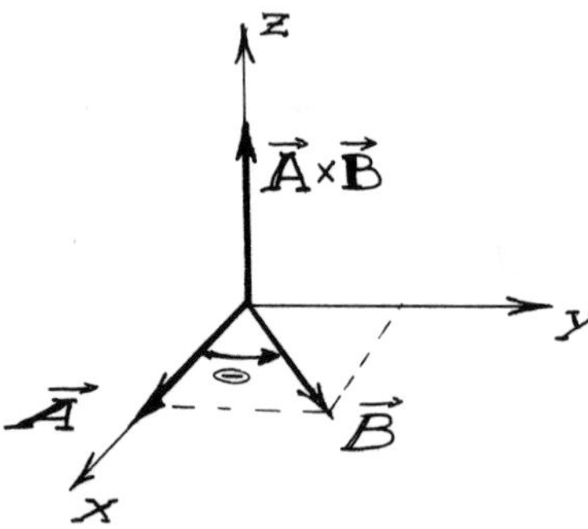

Having defined the direction **A** × **B** we now find that the **order** of the **multiplication** is important. In fact, using our right hand rule we see that an inversion in the order changes the sign but not the magnitude of the cross product.

$$\mathbf{B} \times \mathbf{A} = -\mathbf{A} \times \mathbf{B}$$

and $|\mathbf{C}| = |\mathbf{B} \times \mathbf{A}| = |\mathbf{A} \times \mathbf{B}| =$

$$|\mathbf{A}|\ |\mathbf{B}| \sin \theta$$

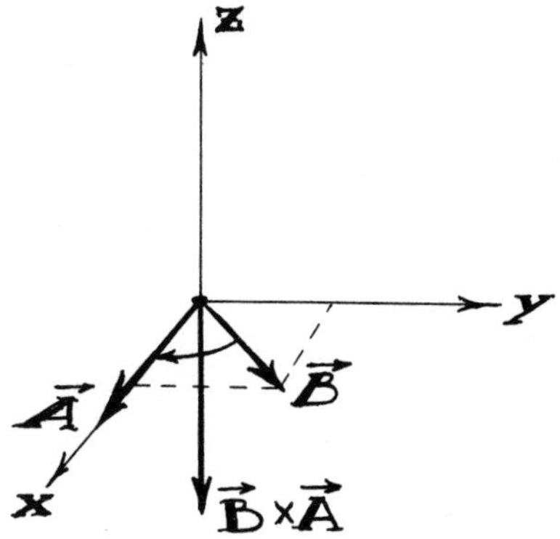

The cross product (or vector product) has many applications in science. In order to illustrate the great utility of the cross product a discussion of mechanical angular momentum will now be presented. This will require some digression from the mathematics and in addition will require the introduction of a few new concepts.

The Angular Momentum of a mass point.

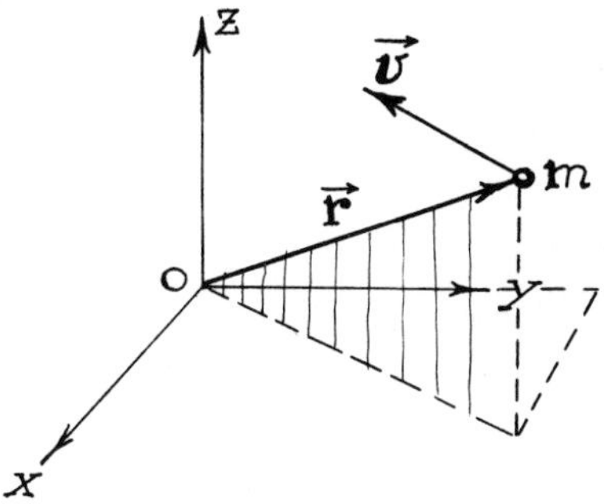

In the figure to the right an idealized system is shown. The entire system consists of a set of coordinates with an origin O and a mass point m moving with a vector velocity **v.** The mass m is located at any given instant in time by a position vector **r.**

The vector velocity **v** is related to the position vector **r.** To see this consider the position of m at two different times t_1 and t_2 with t_2 being the later time. At t_1 the position of m is given by $\mathbf{r}_1$. At t_2 the position of m is given by $\mathbf{r}_2$.

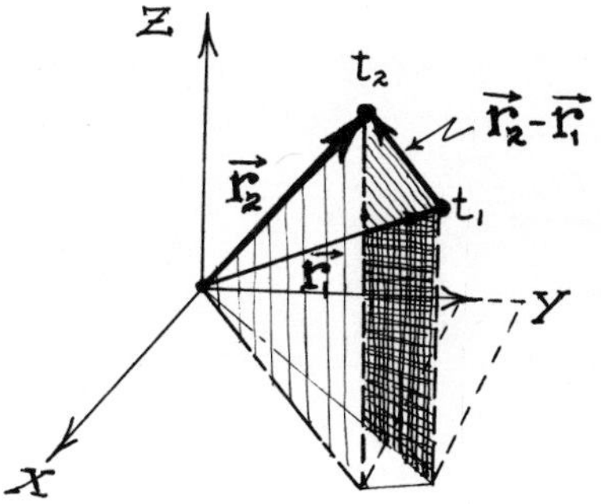

Thus in the time interval $(t_2 - t_1)$ the particle has moved the vector distance given by $\mathbf{r}_2 - \mathbf{r}_1$.

If the time interval $(t_2 - t_1)$ is sufficiently small so that $(\mathbf{r}_2 - \mathbf{r}_1)$ represents the path taken by the point mass m then the vector velocity average is defined by the ratio

$$\mathbf{v} = \frac{(\mathbf{r}_2 - \mathbf{r}_1)}{(t_2 - t_1)}$$

Several things should be noted about this result.

1. The average velocity defined in this manner is that velocity associated with a time intermediate between t_2 and t_1. Say at a time

$$\frac{1}{2}(t_2 + t_1)$$

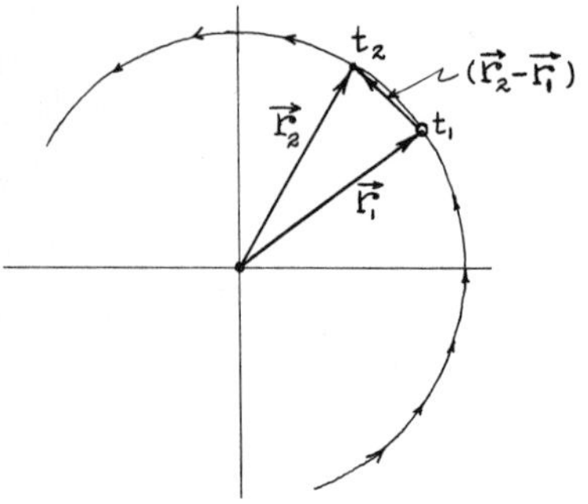

2. In general the time interval $(t_2 - t_1)$ should be very short. To see this consider a point rotating at a constant angular velocity on the circumference of a circle.

If $(\mathbf{r}_2 - \mathbf{r}_1)$ the cord of the arc subtended between t_2 and t_1 is to approximate the path between t_2 and t_1, the cord $(\mathbf{r}_2 - \mathbf{r}_1)$ must be approximately equal to the arc subtended between 1 and 2.

This will only be the case when the angle between $\mathbf{r}_2$ and $\mathbf{r}_1$ is very small.

It should then be apparent that by taking t_2 very close to t_1 the conditions of approximation can be satisfied.

The vector *linear* momentum "**p**" of a mass point is by definition the scalar mass m times the vector velocity **v**;

$$\mathbf{p} = m\mathbf{v}$$

The **vector angular momentum L** relative to O is *defined* as the cross product of the position vector **r** and the linear momentum **p**;

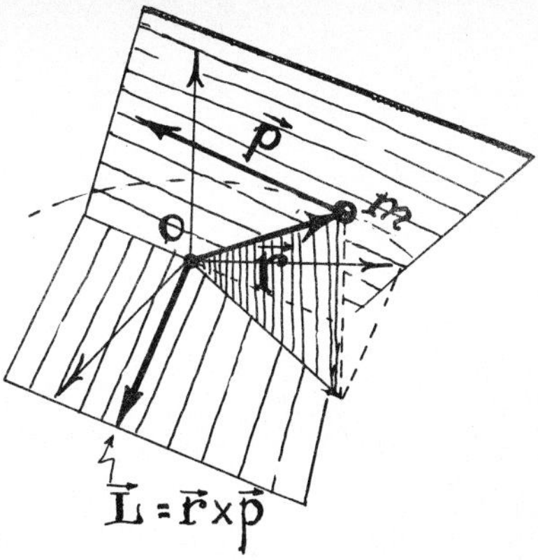

$$\mathbf{L} = \mathbf{r} \times \mathbf{p} = \mathbf{r} \times (m\mathbf{v})$$

Notice in particular that **L** is defined at a specific time t and is *always* defined relative to a specified origin O.

If the position **r** and the velocity **v** vary in direction *and or* magnitude then the angular momentum **L** will vary in direction and magnitude.

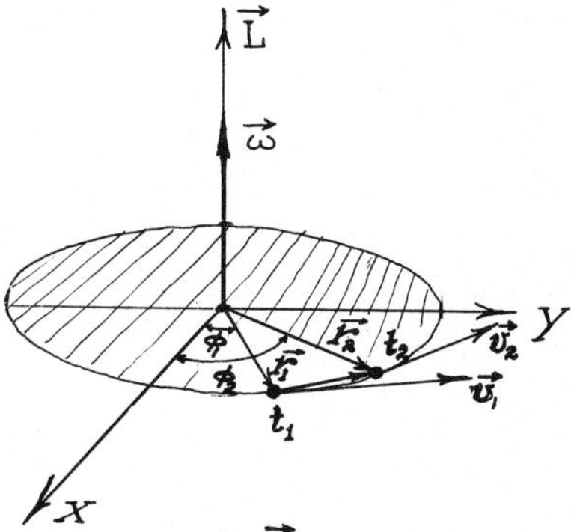

In order that this be more apparent let us again take the example of the point mass moving on the circumference of a circle. Also assume that the magnitude of the velocity of the point is constant.

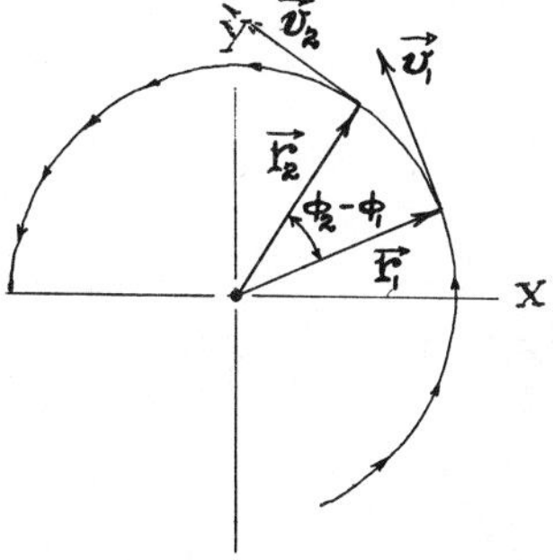

This means that the velocity on the circumference (called the **tangential velocity**) changes only its *direction* in time. Therefore the magnitude of the velocity at t_1 is the same as the magnitude of the velocity at t_2. However the directions of the velocity vectors at t_1 and t_2 are different.

Using the diagrams to the right we can define an additional kinematic vector known as the **angular velocity** ω.

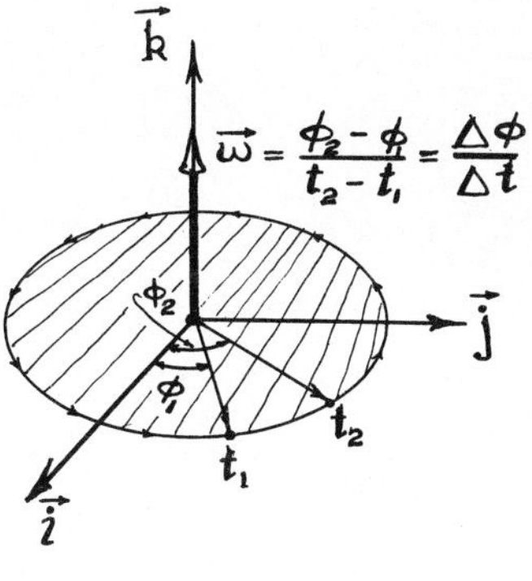

The angular velocity has a *magnitude* equal to the change in angle divided by the change in time

$$|\omega| = \frac{(\phi_2 - \phi_1)}{(t_2 - t_1)}$$

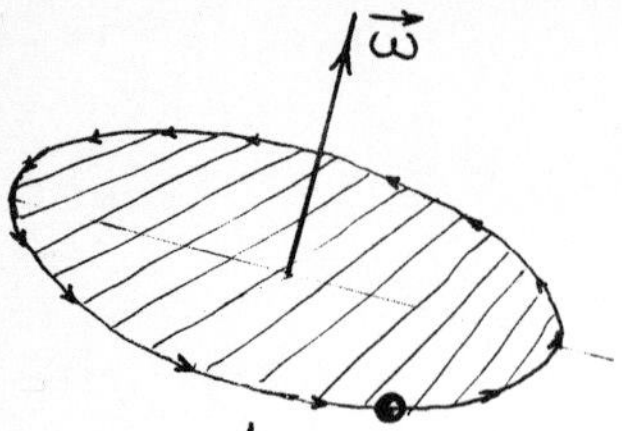

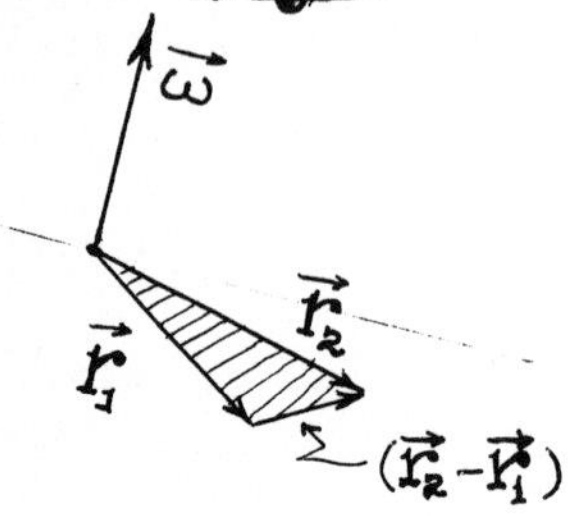

The direction of the angular velocity vector is perpendicular to the plane of $\mathbf{r}_2$ and $\mathbf{r}_1$, and for circular motion ω is perpendicular to the plane of the circle. The direction is the same as the direction of the vector $\mathbf{r}_1 \times \mathbf{r}_2$.

For the circular problem we find that when $\Delta\phi$ is small the length of the cord is approximately equal to the length of the arc subtended by $\mathbf{r}_1$ and $\mathbf{r}_2$. Therefore, if $(\phi_2 - \phi_1)$ is small

$$|\mathbf{r}_2 - \mathbf{r}_1| \simeq (\phi_2 - \phi_1) \cdot |\mathbf{r}_1|$$

Since $(\phi_2 - \phi_1)$ has been defined in terms of ω we can write

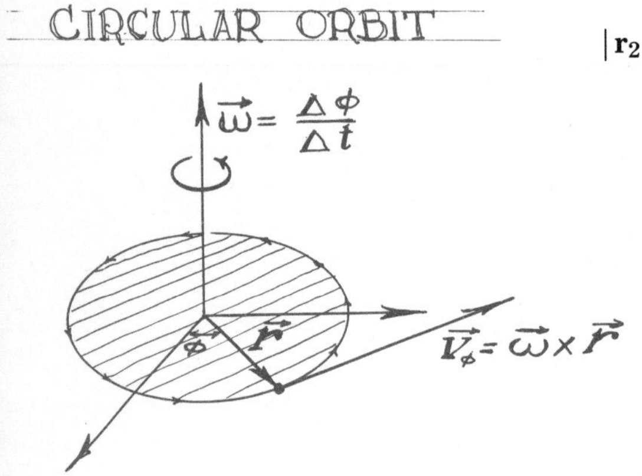

$$|\mathbf{r}_2 - \mathbf{r}_1| \simeq |\omega|\ (t_2 - t_1) \cdot |\mathbf{r}_1|$$

The tangential velocity has been defined, and the magnitude is obtained by dividing by $(t_2 - t_1)$.

$$|\mathbf{v}_t| \simeq \frac{|\mathbf{r}_2 - \mathbf{r}_1|}{t_2 - t_1} = |\omega| \cdot |\mathbf{r}|$$

and in general

$$\mathbf{v}_t = \omega \times \mathbf{r}$$

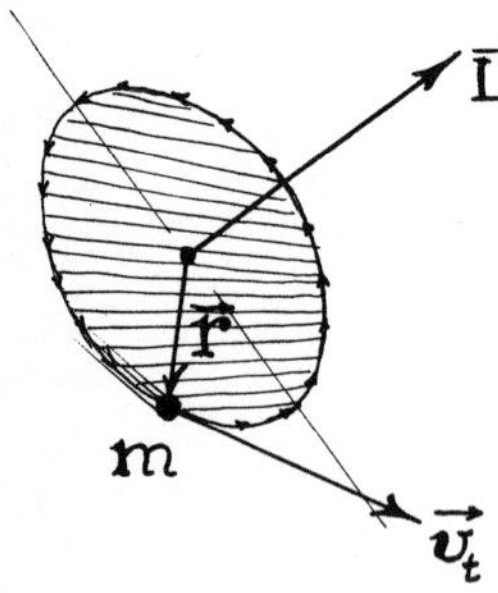

As an exercise the reader should show that in the case of circular motion the direction of ω is uniquely defined in this equation if $\mathbf{v}_t$ and $\mathbf{r}$ are given. Hint, cross $\mathbf{r}$ into $\mathbf{v}_t$ and use the characteristic of circular motion that $\mathbf{r}$ and $\mathbf{v}_t$ are perpendicular, then $\omega = \mathbf{r} \times \mathbf{v}_t/|\mathbf{r}|^2$.

In general

$$\mathbf{L} = \mathbf{r} \times (m\mathbf{v})$$

For circular motion

$$\mathbf{v} = \mathbf{v}_t = \omega \times \mathbf{r}$$

Thus

$$\mathbf{L} = \mathbf{r} \times (m\omega \times \mathbf{r})$$

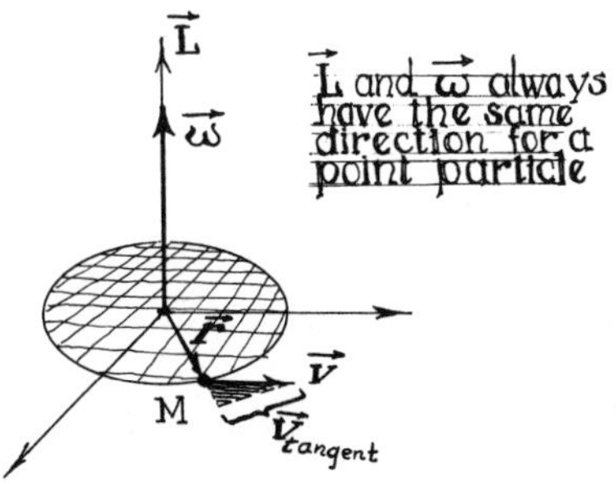

Because ω and $\mathbf{r}$ are mutually perpendicular

$$\mathbf{L} = mr^2\,\omega$$

in other words L and ω are parallel vectors.

It is interesting to note that this result is true for point masses only. When the problem relates to a distributed mass such as a rigid body (a top, gyroscope, etc.) the instantaneous angular momentum **L** and the instantaneous angular velocity ω are not necessarily parallel.

The Triple Scalar Product

Multiple products of more than two vectors can be created by various combinations of the elementary operations of the scalar product and the vector product.

One of the well known triple products is that known as **the triple scalar product.** This particular form arises when a vector product of two vectors is combined with a third vector by means of a scalar product.

Consider the vector product between two vectors **B** and **C,**

$$\mathbf{B} \times \mathbf{C}$$

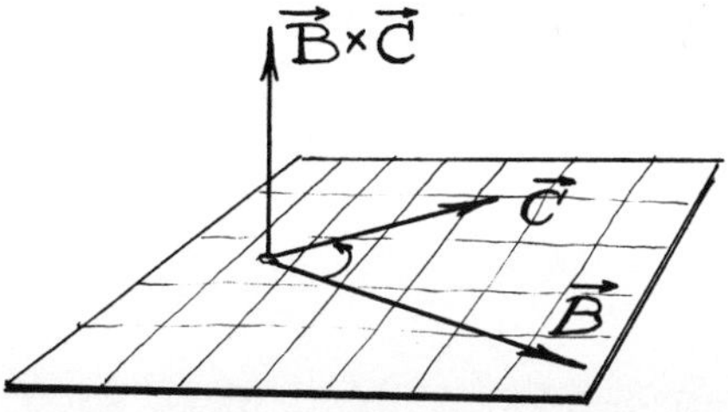

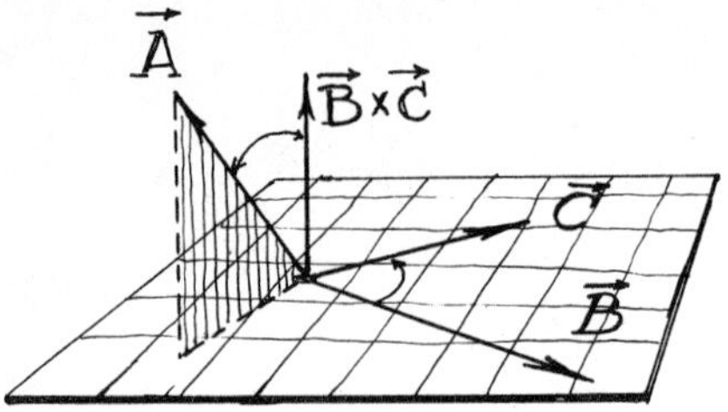

The vector **B** × **C** is perpendicular to the plane of **B** and **C.** Now take the scalar product of (**B** × **C**) with a **third vector A,**

$$\mathbf{A}\cdot(\mathbf{B}\times\mathbf{C})$$

The **triple scalar product** has several interesting properties. First the order can be cycled without changing the sign.

$$\mathbf{A}\cdot(\mathbf{B}\times\mathbf{C}) = \mathbf{C}\cdot(\mathbf{A}\times\mathbf{B}) = \mathbf{B}\cdot(\mathbf{C}\times\mathbf{A})$$

If the order of cycling is changed the product changes sign

$$\mathbf{A}\cdot(\mathbf{B}\times\mathbf{C}) = -\mathbf{A}\cdot(\mathbf{C}\times\mathbf{B}) = -\mathbf{B}\cdot(\mathbf{A}\times\mathbf{C}) = -\mathbf{C}\cdot(\mathbf{B}\times\mathbf{A})$$

The second property is geometrical. If **A, B,** and **C** form the edges of a **parallelepiped,** the magnitude of the triple scalar product is the **volume** of that parallelepiped.

Let θ be the angle between (**B** × **C**) and **A.**

$\mathbf{B}\times\mathbf{C} = |\mathbf{B}|\ |\mathbf{C}|\ \sin \measuredangle_{\mathrm{B}}^{\mathrm{C}}$ is the area of the parallelogram defined by **B** and **C.**

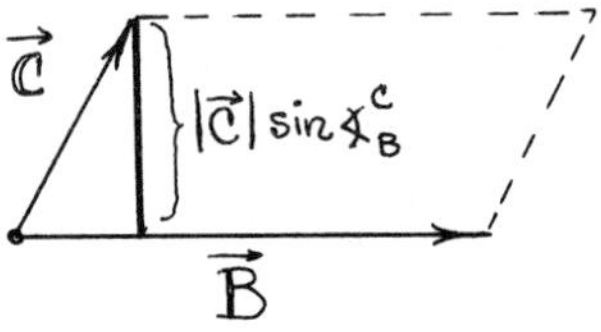

$$\text{Area} = \text{base} \times \text{altitude}$$
$$= |\mathbf{B}|\ |\mathbf{C}|\ \sin \measuredangle_{\mathrm{B}}^{\mathrm{C}}$$

Now the projection of **A** on a line perpendicular to the face **B** × **C** is the height of the parallelepiped.

$$\begin{aligned}\text{Volume} &= |\mathbf{A}|\cos\theta\cdot(\text{base area})\\ &= |\mathbf{A}|\cos\theta\,|(\mathbf{B}\times\mathbf{C})|\\ &= \mathbf{A}\cdot(\mathbf{B}\times\mathbf{C})\end{aligned}$$

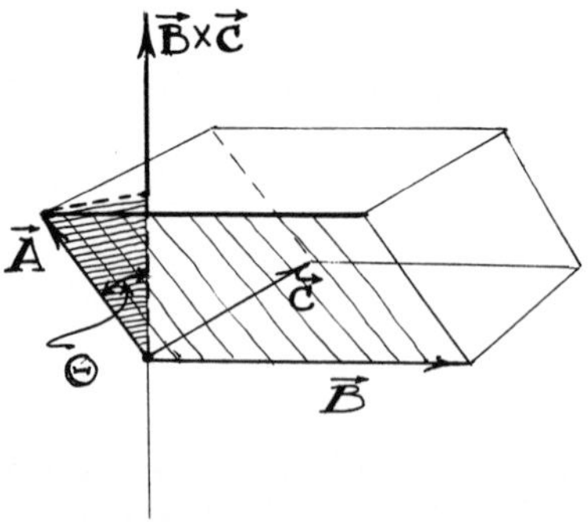

The Triple Vector Product

The vector **B** × **C** is perpendicular to the plane of **B** and **C.**

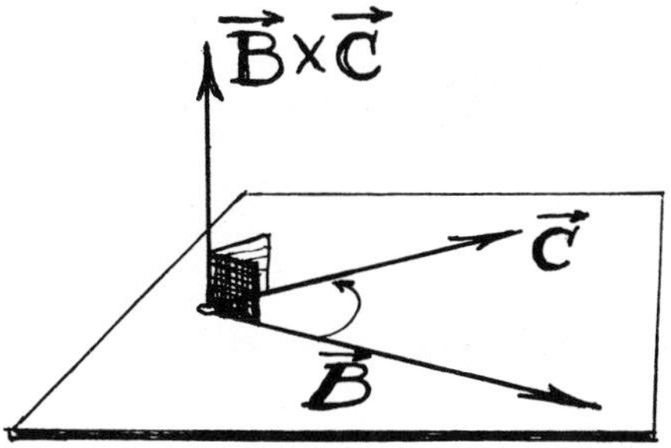

The plane of **B** and **C** contains all lines and vectors perpendicular to the vector **B** × **C** at the point P. The vector **A** × (**B** × **C**) therefore must lie in the plane formed by **B** and **C.**

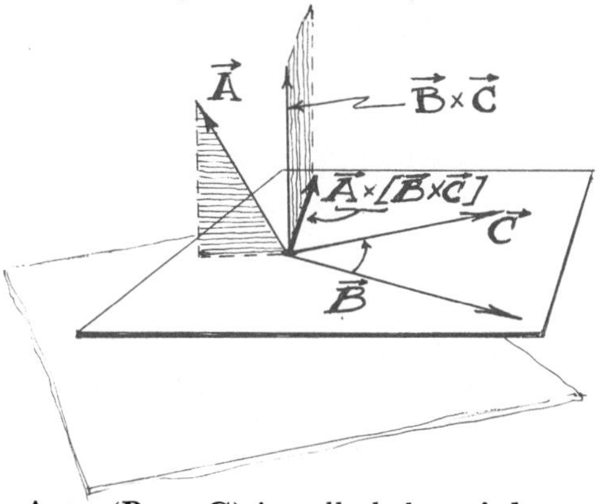

The vector **A** × (**B** × **C**) is called the **triple vector product** of **A, B,** and **C.**

In order to form a non zero vector **B** × **C; B** and **C** must NOT be parallel or anti-parallel. Since **A** × (**B** × **C**) lies in the plane formed by **B** and **C** the triple vector product can be expressed as a sum of two vectors which are parallel or anti-parallel to **B** and **C** respectively.

Another way of saying this is to state that **B** and **C** are **linearly independent** (not parallel or anti-parallel) and form a **subspace** (in this case a plane).

E. Resolution Along a Complete Set of Base Vectors

Introduction

Theorem: A space can be defined in terms of a linearly independent set of vectors and any arbitrary vector *in this space* can be expanded in terms of its projections upon the linearly independent set of vectors.

A set of vectors $\mathbf{F}_1, \mathbf{F}_2 \ldots \mathbf{F}_N$ is said to be **linearly independent** if the relation

$$\sum_{n=1}^{N} a_n \mathbf{F}_n = O$$

is true *if* and *only if* all of the $a_n = O$.

To illustrate this point consider the two vectors **B** and **C** and the scalars b and c. If

$$b\mathbf{B} + c\mathbf{C} = O$$

then, if b and c are **not zero, B** is parallel or anti-parallel to **C.**

$$b\mathbf{B} = -c\mathbf{C}$$

Therefore when **B** and **C** are specified to be linearly independent and are not parallel (or anti-parallel) the only possible solution to $b\mathbf{B} + c\mathbf{C} = O$ is that

$$b = c = O$$

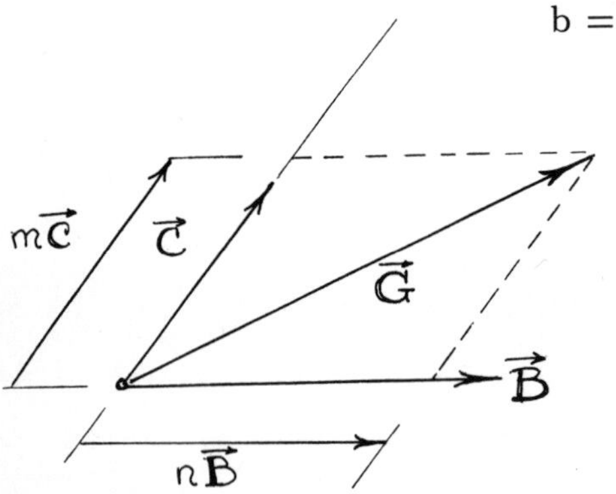

Under this last condition **B** and **C** are said to be **linearly independent.**

Now regard an arbitrary vector **G** in the plane (or subspace) defined by **B** and **C.**

The projection[2] of **G** on **B** and **C** is performed just as we projected a point P on the general coordinate axes in Chapter 1.

To perform this expansion construct a vector along **B** and a vector along **C** such that

$$n\mathbf{B} + m\mathbf{C} = \mathbf{G}$$

This operation is an elementary example of the **expansion theorem** given previously.

After this digression we can see that because $\mathbf{A} \times (\mathbf{B} \times \mathbf{C})$ lies in the plane (subspace) defined by **B** and **C**, the triple vector product can be expanded in terms of the vectors **B** and **C**.

We shall quote the result without proof at this point. The proof can be performed when we have established the base vectors; however, it is a straight-forward operation but tedious.[3]

The expansion of $\mathbf{A} \times (\mathbf{B} \times \mathbf{C})$ in terms of **B** and **C** is called the BAC **minus** CAB formula,

$$\begin{aligned} \mathbf{A} \times (\mathbf{B} \times \mathbf{C}) &= \mathbf{B}\,(\mathbf{A} \cdot \mathbf{C}) - \mathbf{C}\,(\mathbf{A} \cdot \mathbf{B}) \\ &= n\mathbf{B} - m\mathbf{C} \end{aligned}$$

$$\begin{aligned} n &= (\mathbf{A} \cdot \mathbf{C}) \\ m &= (\mathbf{A} \cdot \mathbf{B}) \end{aligned}$$

A note of warning should be made concerning the brackets about $\mathbf{B} \times \mathbf{C}$ in

$$\mathbf{A} \times (\mathbf{B} \times \mathbf{C})$$

[2] Note that this is not a projection as described at the beginning of this chapter. This is a projection obtained by passing lines through the terminus of **G** parallel to **B** and **C**. The intersections of these lines with extensions of **B** and **C** determine the lengths of the projections (See Chapter 1).

[3] We can verify this proof up to a constant multiplier in the following fashion. Assume $\mathbf{A} \times (\mathbf{B} \times \mathbf{C}) = n\mathbf{B} - m\mathbf{C}$. Take the scalar product of this equation with **A** and utilize the properties of the triple scalar product.

$$\mathbf{A} \cdot \mathbf{A} \times (\mathbf{B} \times \mathbf{C}) = (\mathbf{B} \times \mathbf{C}) \cdot (\mathbf{A} \times \mathbf{A}) = 0 = n(\mathbf{A} \cdot \mathbf{B}) - m(\mathbf{A} \cdot \mathbf{C})$$

Then

$$n = \mathbf{A} \cdot \mathbf{C} \text{ and } m = \mathbf{A} \cdot \mathbf{B} \text{ is a solution.}$$

These brackets specify that the **cross product** of **B** and **C** must be taken **before** the cross product is taken with **A.**

It can be shown for instance that

$$\mathbf{A} \times (\mathbf{B} \times \mathbf{C}) \neq (\mathbf{A} \times \mathbf{B}) \times \mathbf{C}$$

Base Vectors

Let **A, B,** and **C** be three non-coplanar vectors issuing from a common point O.

If **A, B,** and **C** are **non-coplanar** vectors, they are **linearly independent,** and any arbitrary vector **F** at O can be represented as the resultant of the sum of three *vectors* which are parallel to **A, B,** and **C** respectively. In otherwords, it is possible to write

$$\mathbf{F} = a\mathbf{A} + b\mathbf{B} + c\mathbf{C}$$

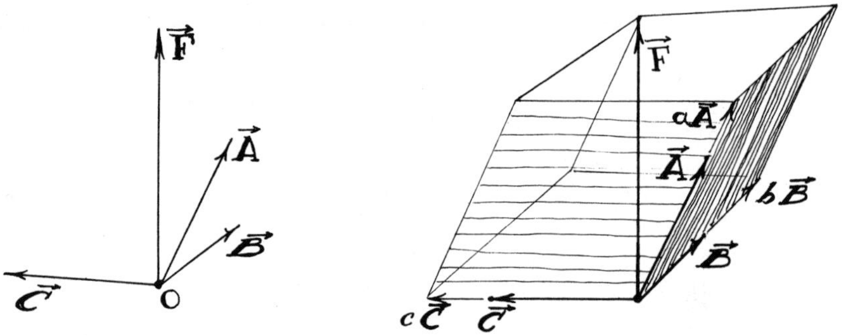

To find a**A** + b**B** + c**C** we must construct a parallelepiped with edges a**A,** b**B,** and c**C** as shown above.

It is apparent that a suitable choice of a, b, and c will give a representation of **F** provided **A, B,** and **C** do not lie in a plane.

One can stipulate that **A, B,** and **C** are non-coplanar (or linearly independent) by requiring that $\mathbf{A} \cdot (\mathbf{B} \times \mathbf{C}) \neq 0$.

To construct the parallelepiped we pass planes through the terminal end of **F,** parallel to the three planes defined by (**A, B**); (**B, C**); and (**C, A**). The intersection of the plane (**B, C**) with the extension of **A** provides the length of the vector a**A.**

Orthogonal bases of arbitrary length.

If the three vectors **A, B** and **C** at O are **mutually orthogonal,*** the resolution of an arbitrary vector **F** along them becomes a much more straightforward procedure in terms of the vector operations which we have defined.

Consider the three mutually orthogonal vectors **A, B,** and **C** and an arbitrary vector **F.**

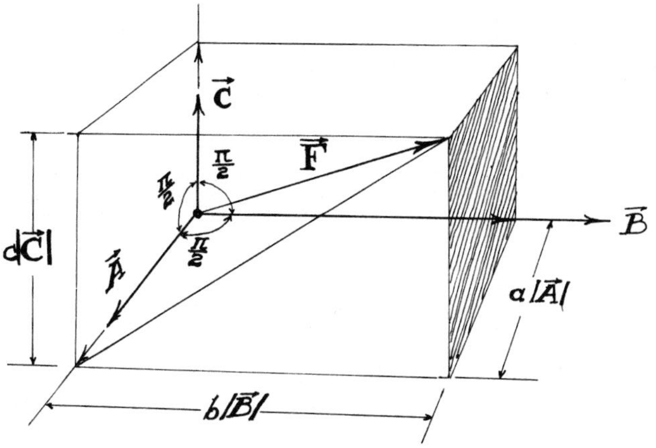

The parallelepiped in this case is a rectangular parallelepiped. Under these conditions of orthogonality the numbers, a, b, and c can be obtained by calculating ***the projection*** of **F** upon **A, B,** and **C** respectively.

Regard the triangle OPQ shown in the diagram.

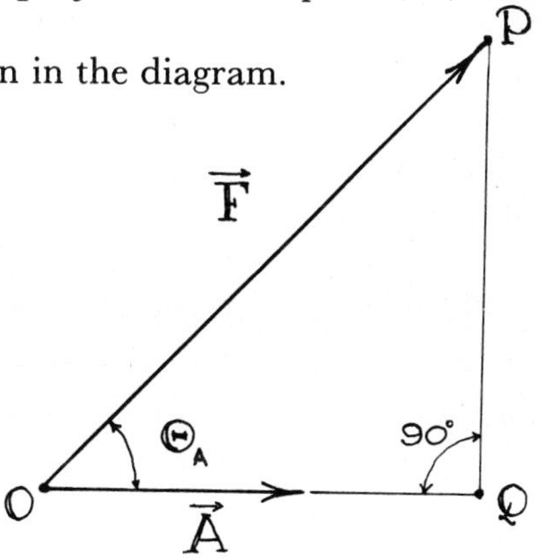

* One vector is orthogonal to a second vector when the two vectors are mutually perpendicular (i.e., the angle between them is 90°).

Previously we defined the projection of one vector upon another in terms of the **dot or scalar product.**

$$\mathbf{F} \cdot \mathbf{A} = |\mathbf{F}|\ |\mathbf{A}| \cos\theta_A$$

$$a\mathbf{A} = |\mathbf{F}| \cos\theta_A \frac{\mathbf{A}}{|\mathbf{A}|}$$

Thus

$$a\mathbf{A} = \frac{(\mathbf{F} \cdot \mathbf{A})}{|\mathbf{A}|^2} \mathbf{A}$$

In the same manner,

$$b\mathbf{B} = \frac{(\mathbf{F} \cdot \mathbf{B})}{|\mathbf{B}|^2} \mathbf{B}$$

and

$$c\mathbf{C} = \frac{(\mathbf{F} \cdot \mathbf{C})}{|\mathbf{C}|^2} \mathbf{C}$$

The vector **F** resolved along **A, B,** and **C** can now be written

$$\mathbf{F} = \left(\mathbf{F} \cdot \frac{\mathbf{A}}{|\mathbf{A}|}\right) \frac{\mathbf{A}}{|\mathbf{A}|} + \left(\mathbf{F} \cdot \frac{\mathbf{B}}{|\mathbf{B}|}\right) \frac{\mathbf{B}}{|\mathbf{B}|} + \left(\mathbf{F} \cdot \frac{\mathbf{C}}{|\mathbf{C}|}\right) \frac{\mathbf{C}}{|\mathbf{C}|}$$

It is important to see at this point that the base vectors **A, B,** and **C** have been written in a form which provides **three unit vectors** $\frac{\mathbf{A}}{|\mathbf{A}|}$, $\frac{\mathbf{B}}{|\mathbf{B}|}$, and $\frac{\mathbf{C}}{|\mathbf{C}|}$. These unit vectors will be denoted by the symbol ϵ.

Let

$$\epsilon_A = \frac{\mathbf{A}}{|\mathbf{A}|}$$

and

$$\epsilon_B = \frac{\mathbf{B}}{|\mathbf{B}|}$$

$$\epsilon_C = \frac{\mathbf{C}}{|\mathbf{C}|}$$

then

$$\mathbf{F} = (\mathbf{F} \cdot \epsilon_A)\epsilon_A + (\mathbf{F} \cdot \epsilon_B)\epsilon_B + (\mathbf{F} \cdot \epsilon_C)\epsilon_C$$

This process of **normalization** (converting the bases to unit bases) leads us naturally to the next section which deals with the Cartesian unit base vectors.

The Cartesian Bases.

In the first chapter concerning geometry we defined the three dimensional **Euclidean space** with the coordinate axes x, y, and z and an origin O.

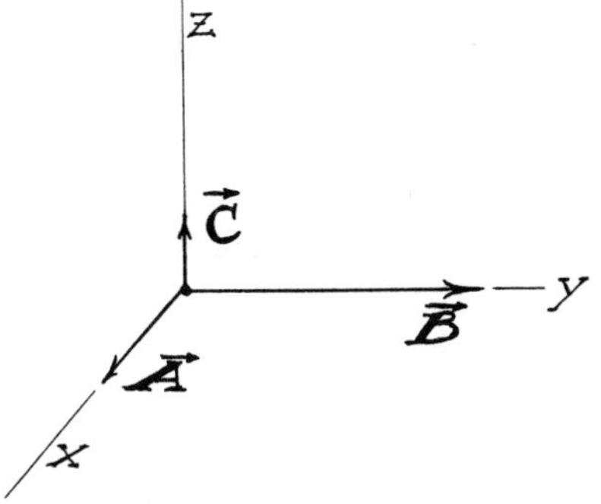

We now set up the three base vectors **A, B,** and **C** along the **positive** x, y, and z axes respectively.

The order of the base vectors is taken to provide a **right handed system.**

The unit base vectors for the Cartesian space have standard symbols. Assuming that **A** lies along the positive x axis; **B** along the positive y axis; and **C** along the z axis; the unit base vectors are defined as

$$\mathbf{i} = \frac{\mathbf{A}}{|\mathbf{A}|} = \epsilon_A$$

$$\mathbf{j} = \frac{\mathbf{B}}{|\mathbf{B}|} = \epsilon_B$$

and

$$\mathbf{k} = \frac{\mathbf{C}}{|\mathbf{C}|} = \epsilon_C$$

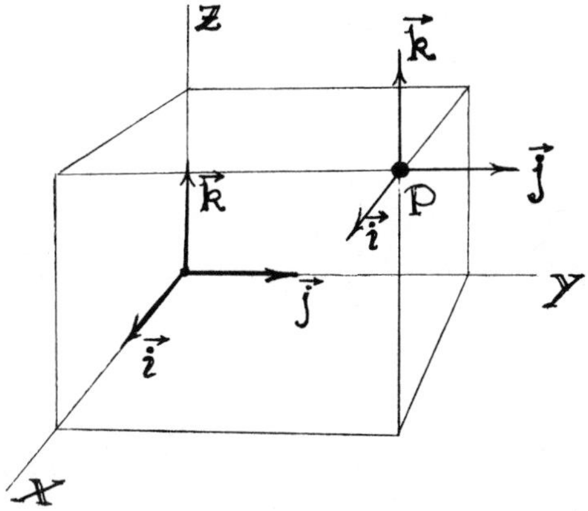

Actually **six quantities** are required to specify a vector in the Cartesian space. We assume that there is a set of Cartesian base vectors associated with every point in E_3. Thus to specify a vector at a point P we require,

1. The position vector of the point P (3 numbers).
2. The components of the vector along the bases, **i, j,** and **k.**

Using the **i, j, k** notation for the three unit base vectors in E_3 the arbitrary vector **F** is written.

$$\mathbf{F} = (\mathbf{F} \cdot \mathbf{i})\,\mathbf{i} + (\mathbf{F} \cdot \mathbf{j})\,\mathbf{j} + (\mathbf{F} \cdot \mathbf{k})\,\mathbf{k}$$

The scalar coefficients $(\mathbf{F} \cdot \mathbf{i})$, $(\mathbf{F} \cdot \mathbf{j})$, and $(\mathbf{F} \cdot \mathbf{k})$ are called the **Cartesian components** of the vector **F** and are written as

$$\begin{aligned} F_x &= (\mathbf{F} \cdot \mathbf{i}) = |\mathbf{F}| \cos\alpha \\ F_y &= (\mathbf{F} \cdot \mathbf{j}) = |\mathbf{F}| \cos\beta \\ F_z &= (\mathbf{F} \cdot \mathbf{k}) = |\mathbf{F}| \cos\gamma \end{aligned}$$

$\cos\alpha$, $\cos\beta$ and $\cos\gamma$ are the **direction cosines** of the vector **F** relative to the x, y, and z axes. The components F_x, F_y, and F_z plus the direction cosines are illustrated in the figure below.

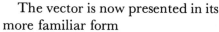

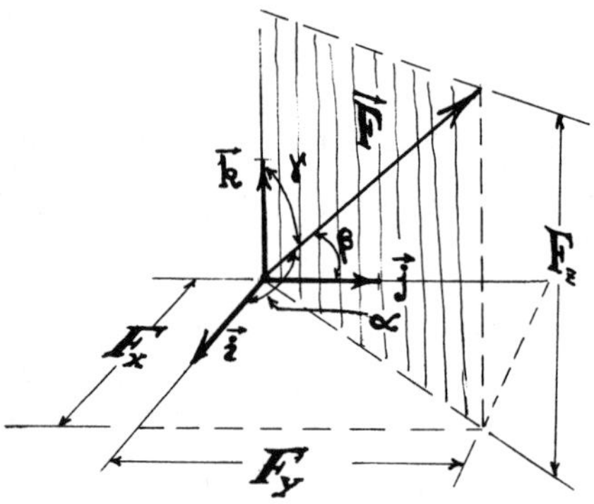

The vector is now presented in its more familiar form

$$\mathbf{F} = F_x\,\mathbf{i} + F_y\,\mathbf{j} + F_z\,\mathbf{k}$$

The notation can be contracted by representing F_x as F_1, F_y as F_2, F_z as F_3; in addition we can write ϵ_1, ϵ_2, and ϵ_3 in place of **i, j, k.** Then

$$\mathbf{F} = F_1\epsilon_1 + F_2\epsilon_2 + F_3\epsilon_3 = \sum_{n=1}^{n=3} F_n\epsilon_n$$

This notation conforms to the more general symbolism which was used in the first chapter in the section concerning Matrices.

In fact if we utilize our **summation convention** for repeated indices the vector can be written as

$$\mathbf{F} = F_n\epsilon_n$$

The Addition of Resolved Vectors

Two vectors **F** and **G** can be added to form a third vector **R**; **F** + **G** = **R.** Previously we performed the addition by the parallelogram method. At that time we saw that the parallelogram method was equivalent to a construction in which the tail of one vector is connected to the head of the other and the resultant is obtained by drawing a vector from the tail of the sum to the head of the sum.

Addition of resolved vectors is an algebraic operation whereas the addition of unresolved vectors is a graphical problem.

> When two vectors **F** and **G** are added, the component of a given base vector for the resultant vector **R** is the sum of the components of **F** and **G** corresponding to the same base vector.

The diagram to the right illustrates in two dimensions how components of like base vectors are summed.

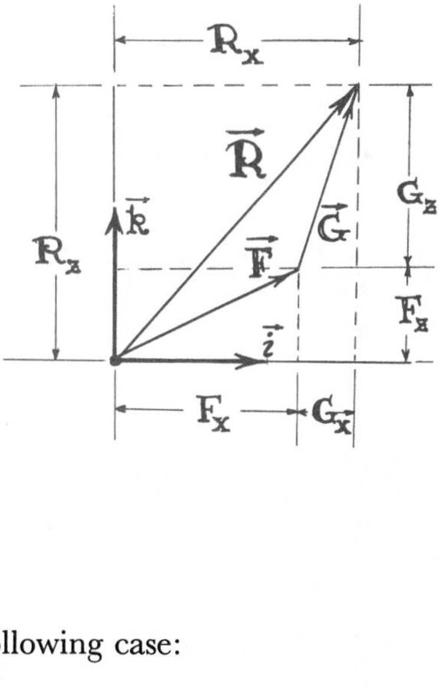

If $\mathbf{R} = \mathbf{F} + \mathbf{G}$

Where $\mathbf{F} = F_x\mathbf{i} + F_y\mathbf{j} + F_z\mathbf{k}$

and $\mathbf{G} = G_x\mathbf{i} + G_y\mathbf{j} + G_z\mathbf{k}$

Then

$$\mathbf{F} + \mathbf{G} = (F_x + G_x)\mathbf{i} + (F_y + G_y)\mathbf{j} + (F_z + G_z)\mathbf{k}$$

$$\mathbf{R} = R_x\,\mathbf{i} + R_y\,\mathbf{j} + R_z\mathbf{k}$$

and $R_x = F_x + G_x$

$R_y = F_y + G_y$

$R_z = F_z + G_z$

As a numerical example consider the following case:

Let $\mathbf{F} = 1\mathbf{i} - 3\mathbf{j} + 10\mathbf{k}$

and $\mathbf{G} = \frac{1}{2}\mathbf{i} + 5\mathbf{j} - \left(\frac{7}{2}\right)\mathbf{k}$

The sum **F** + **G** is then given by

$$\mathbf{R} = \mathbf{F} + \mathbf{G} = \frac{3}{2}\mathbf{i} + 2\mathbf{j} + \left(\frac{13}{2}\right)\mathbf{k}$$

Because subtraction is a special case of addition we can write

$$\mathbf{F} - \mathbf{G} = (F_x - G_x)\mathbf{i} + (F_y - G_y)\mathbf{j} + (F_z - G_z)\mathbf{k}$$

Products of Resolved Vectors

The dot or scalar product of two resolved vectors.

When one considers the dot product of two vectors **F** and **G** which have been resolved into components along the base vectors, **i, j,** and **k** the final result is determined conveniently by the *scalar products* of the unit base vectors.

$$\begin{array}{lll} \mathbf{i}\cdot\mathbf{i} = 1 & \mathbf{j}\cdot\mathbf{i} = 0 & \mathbf{k}\cdot\mathbf{i} = 0 \\ \mathbf{i}\cdot\mathbf{j} = 0 & \mathbf{j}\cdot\mathbf{j} = 1 & \mathbf{k}\cdot\mathbf{j} = 0 \\ \mathbf{i}\cdot\mathbf{k} = 0 & \mathbf{j}\cdot\mathbf{k} = 0 & \mathbf{k}\cdot\mathbf{k} = 1 \end{array}$$

The orthogonality of the bases leads to convenient relations for scalar products.

$$\begin{aligned} \mathbf{F}\cdot\mathbf{G} &= (F_x\mathbf{i} + F_y\mathbf{j} + F_z\mathbf{k})\cdot(G_x\mathbf{i} + G_y\mathbf{j} + G_z\mathbf{k}) \\ &= F_xG_x(\mathbf{i}\cdot\mathbf{i}) + F_xG_y(\mathbf{i}\cdot\mathbf{j}) + F_xG_z(\mathbf{i}\cdot\mathbf{k}) \\ &\quad + F_yG_x(\mathbf{j}\cdot\mathbf{i}) + F_yG_y(\mathbf{j}\cdot\mathbf{j}) + F_yG_z(\mathbf{j}\cdot\mathbf{k}) \\ &\quad + F_zG_x(\mathbf{k}\cdot\mathbf{i}) + F_zG_y(\mathbf{k}\cdot\mathbf{j}) + F_zG_z(\mathbf{k}\cdot\mathbf{k}) \end{aligned}$$

and

$$\mathbf{F}\cdot\mathbf{G} = F_xG_x + F_yG_y + F_zG_z$$

note: $|\mathbf{F}| = \sqrt{\mathbf{F}\cdot\mathbf{F}} = \sqrt{F_x^2 + F_y^2 + F_z^2}$

It is interesting to observe the compactness achieved in such a development when the more generalized notation is used.

If **F** and **G** are written

$$\mathbf{F} = \sum_{m=1}^{3} F_m \epsilon_m$$

and

$$G = \sum_{n=1}^{3} G_n \epsilon_n$$

Then

$$\mathbf{F} \cdot \mathbf{G} = \{\sum_{m=1}^{3} F_m \epsilon_m\} \cdot \{\sum_{n=1}^{3} G_n \epsilon_n\}$$

$$= \sum_{m=1}^{3} \sum_{n=1}^{3} F_m G_n (\epsilon_m \cdot \epsilon_n)$$

Now in the case of orthonormal bases

$$\epsilon_m \cdot \epsilon_n = \delta_{mn}$$

where

$$\delta_{mn} = 1 \text{ if } m = n$$
$$= 0 \text{ if } m \neq n$$

Our scalar product becomes[4]

$$\mathbf{F} \cdot \mathbf{G} = \sum_{m=1}^{3} \sum_{n=1}^{3} F_m G_n \delta_{mn}$$

$$= \sum_{m=1}^{3} F_m G_m$$

$$= F_1 G_1 + F_2 G_2 + F_3 G_3$$

[4] This particularly simple expansion as the sum of the products of like components is characteristic only of the orthogonal systems of base vectors.

To illustrate a typical scalar product the vectors of the example of the previous section can be used again,

$$F = \mathbf{i} - 3\mathbf{j} + 10\mathbf{k}$$
$$G = \frac{1}{2}\mathbf{i} + 5\mathbf{j} - \left(\frac{7}{2}\right)\mathbf{k}$$

Thus

$$\mathbf{F} \cdot \mathbf{G} = (1 \times 1/2) - (3 \times 5) - (7/2 \times 10) = 49\frac{1}{2}$$

The vector product

Once more the orthonormal property of **i, j, k** provides a convenient result when we consider the **vector product** of two arbitrary vectors **F** and **G.**

$$\begin{aligned}\mathbf{F} \times \mathbf{G} = {} & F_xG_x\,(\mathbf{i} \times \mathbf{i}) + F_xG_y\,(\mathbf{i} \times \mathbf{j}) + F_xG_z\,(\mathbf{i} \times \mathbf{k}) \\ & + F_yG_x\,(\mathbf{j} \times \mathbf{i}) + F_yG_y\,(\mathbf{j} \times \mathbf{j}) + F_yG_z\,(\mathbf{j} \times \mathbf{k}) \\ & + F_zG_x\,(\mathbf{k} \times \mathbf{i}) + F_zG_y\,(\mathbf{k} \times \mathbf{j}) + F_zG_z\,(\mathbf{k} \times \mathbf{k})\end{aligned}$$

Using the Right Hand Convention the *cross products* of the unit base vectors can be determined.

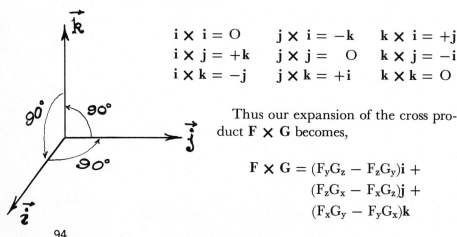

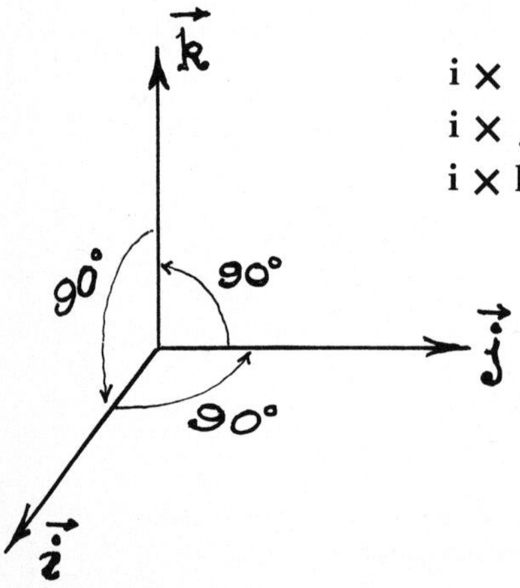

$\mathbf{i} \times \mathbf{i} = 0$	$\mathbf{j} \times \mathbf{i} = -\mathbf{k}$	$\mathbf{k} \times \mathbf{i} = +\mathbf{j}$
$\mathbf{i} \times \mathbf{j} = +\mathbf{k}$	$\mathbf{j} \times \mathbf{j} = 0$	$\mathbf{k} \times \mathbf{j} = -\mathbf{i}$
$\mathbf{i} \times \mathbf{k} = -\mathbf{j}$	$\mathbf{j} \times \mathbf{k} = +\mathbf{i}$	$\mathbf{k} \times \mathbf{k} = 0$

Thus our expansion of the cross product **F × G** becomes,

$$\begin{aligned}\mathbf{F} \times \mathbf{G} = {} & (F_yG_z - F_zG_y)\mathbf{i} + \\ & (F_zG_x - F_xG_z)\mathbf{j} + \\ & (F_xG_y - F_yG_x)\mathbf{k}\end{aligned}$$

The **cross product** expanded in terms of the components of the two vectors involved can be expressed in terms of a determinant.

$$\mathbf{F} \times \mathbf{G} = \begin{vmatrix} \mathbf{i} & \mathbf{j} & \mathbf{k} \\ F_x & F_y & F_z \\ G_x & G_y & G_z \end{vmatrix}$$

1. The base vectors occupy the first row.
2. The components of the leading vector in the product occupy the second row.
3. The components of the last vector in the product occupy the third row.

The proof of the validity of the determinant form can be obtained by expanding the determinant and comparing the result with the form obtained previously.

Once again we can utilize the example vectors to demonstrate a typical calculation employing the cross product.

As before

$$F = \mathbf{i} - 3\mathbf{j} + 10\mathbf{k}$$
$$G = 1/2\mathbf{i} + 5\mathbf{j} - (7/2)\mathbf{k}$$

Then

$$\mathbf{F} \times \mathbf{G} = \begin{vmatrix} \mathbf{i} & \mathbf{j} & \mathbf{k} \\ 1 & -3 & 10 \\ 1/2 & 5 & -7/2 \end{vmatrix}$$

$$\mathbf{F} \times \mathbf{G} = (21/2 - 50)\mathbf{i} + (5 + 7/2)\mathbf{j} + (5 + 3/2)\mathbf{k}$$

$$= -\left(\frac{79}{2}\right)\mathbf{i} + \left(\frac{17}{2}\right)\mathbf{j} + \left(\frac{13}{2}\right)\mathbf{k}$$

We should note in closing this section that the triple scalar product of three arbitrary vectors **A, B,** and **C** assumes a particularly convenient form in terms of the components of the three vectors.

If we wish to compute

$$\mathbf{A} \cdot (\mathbf{B} \times \mathbf{C})$$

then
$$(\mathbf{B} \times \mathbf{C}) = \begin{vmatrix} \mathbf{i} & \mathbf{j} & \mathbf{k} \\ B_x & B_y & B_z \\ C_x & C_y & C_z \end{vmatrix}$$

and
$$\mathbf{A} \cdot (\mathbf{B} \times \mathbf{C}) = \begin{vmatrix} A_x & A_y & A_z \\ B_x & B_y & B_z \\ C_x & C_y & C_z \end{vmatrix}$$

In other words the triple scalar product of three vectors, **A, B,** and **C** is merely the **determinant** of their components. The cycling procedure can now be seen to be equivalent to an interchange of rows. The negative sign which arises when one writes

$$\mathbf{A} \cdot (\mathbf{B} \times \mathbf{C}) = -\mathbf{B} \cdot (\mathbf{A} \times \mathbf{C})$$

can be associated with the sign change which occurs when two rows of the determinant are interchanged.

F. Vector Transformations

In Chapter 1 the term "vector" was used interchangeably with the term "column matrix" (or row matrix). This labeling is permissible because the representation of a point in a specified space can be accomplished by using either the listing of the coordinates of the point in an ordered array (a column matrix) or by representing coordinates with their associated base vectors.

The two representations are equivalent.

The particular representation which one uses depends upon the problem at hand and upon the convenience with which the symbols can be manipulated.

The equivalence of the two representations is illustrated by the fact that they possess the same transformation properties. The vector or linear array of objects is one of that group of mathematical objects called **tensors.**

The order of a tensor depends upon its transformation properties.

Consider the rotations (orthogonal transformations) which we have already discussed.

Under these transformations

1. The Scalar quantity is invariant. The scalar is called a zeroth order tensor.
2. The vector transforms in the following manner:

$$\mathbf{r}' = S \cdot \mathbf{r}$$

or

$$x_j' = \sum_k S_{jk}\, x_k$$

or

$$\begin{bmatrix} x_1' \\ x_2' \end{bmatrix} = \begin{pmatrix} S_{11} & S_{12} \\ S_{21} & S_{22} \end{pmatrix} \begin{bmatrix} x_1 \\ x_2 \end{bmatrix}$$

Objects which transform in this manner are called **first order tensors** or **vectors.**

3. We should also remark that a matrix A transforms as follows:

$$B = S \cdot A \cdot S^{-1}$$

or

$$B_{lm} = \sum_{\text{all } k} \sum_{\text{all } n} S_{lk}\, A_{kn}\, S_{nm}'$$

Objects which transform in the manner shown above are called **second order tensors.**

Our aim is not to provide a complete classification of tensors by their transformation properties but rather to indicate that such a classification exists and that it should be kept in mind.

When a vector is described in terms of its matrix form certain characteristics must be considered. To introduce these characteristics we shall employ a simple example.

As an example, regard the position vector **r** which locates the point (1, 3, 2) in a three dimensional vector space and which can be described by the column matrix

$$\mathbf{r} = \begin{bmatrix} 1 \\ 3 \\ 2 \end{bmatrix}$$

or by the vector form

$$r = \epsilon_1 + 3\epsilon_2 + 2\epsilon_3$$

Notice that the matrix representation suppresses the base vectors; in other words one must understand that the base ϵ_1 is associated with the first position in the column and that ϵ_m is associated with the mth position.

Inner products or **dot products** are taken in the matrix notation in the following manner,

$$\tilde{\mathbf{r}} \cdot \mathbf{r} = [1, 3, 2] \begin{bmatrix} 1 \\ 3 \\ 2 \end{bmatrix} = 14$$

or in the vector notation

$$\mathbf{r} \cdot \mathbf{r} = \sum_{l} \sum_{k} x_l\, x_k\, (\epsilon_l \cdot \epsilon_k) = \sum_{l} x_l^2$$

In many cases (such as that of rotations) the transformation of a vector may be achieved most easily when the vector is represented in its matrix form.

To illustrate this we can consider the rotations as applied to vectors. This procedure seems paradoxical in the following sense.

The rotation matrix can be thought of as **a rotation of the coordinates,**

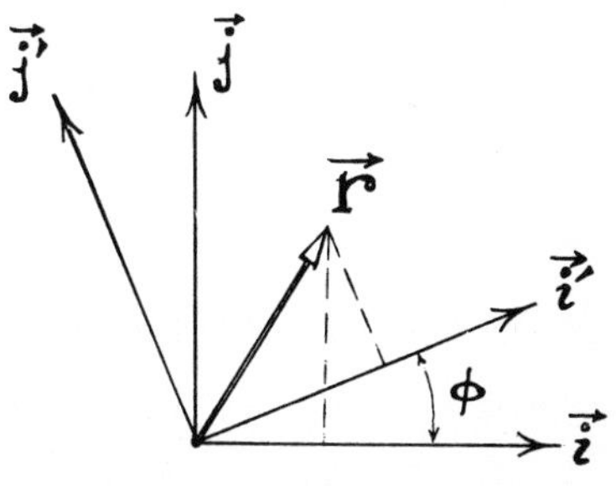

$$\mathbf{r}' = \mathbb{S} \cdot \mathbf{r}$$

In other words the application of $\mathbb{S}$ to $\mathbf{r}$ (described in terms of the bases $\mathbf{i}$ and $\mathbf{j}$)[5] gives the same vector in a new set of coordinates $\mathbf{i}'$ and $\mathbf{j}'$ which make an angle ϕ with respect to the old coordinates. Stipulating this interpretation of $\mathbb{S}$ we consider the vector to be fixed in $\mathbf{i}$ and $\mathbf{j}$ and then obtain its components in system $\mathbf{i}'$, $\mathbf{j}'$ rotated counter clockwise with respect to $\mathbf{i}$ and $\mathbf{j}$.

$$\mathbf{i} \cdot \mathbf{i}' = \cos \phi$$

If the problem deals with a set of moving bases; then the *positions* of the *components* in the final vector must be associated with the *new bases.*

Consider the rotation of the bases

$$\mathbf{r}' = \mathbb{S} \cdot \mathbf{r}$$

or

$$\begin{bmatrix} x_1' \\ x_2' \end{bmatrix} = \begin{pmatrix} S_{11} & S_{12} \\ S_{21} & S_{22} \end{pmatrix} \begin{bmatrix} x_1 \\ x_2 \end{bmatrix}$$

$$\mathbf{r} = x_1\mathbf{i} + x_2\mathbf{j}$$

and

$$\mathbf{r}' = x_1'\,\mathbf{i}' + x_2'\,\mathbf{j}'$$

[5] We use ϵ_1 and $\mathbf{i}$ interchangeably. This alternate use of the different symbols will familiarize the reader with the symbols and their relationships.

In this example the base vectors after transformation are not the same bases as used in the expansion of the vector in the original system.

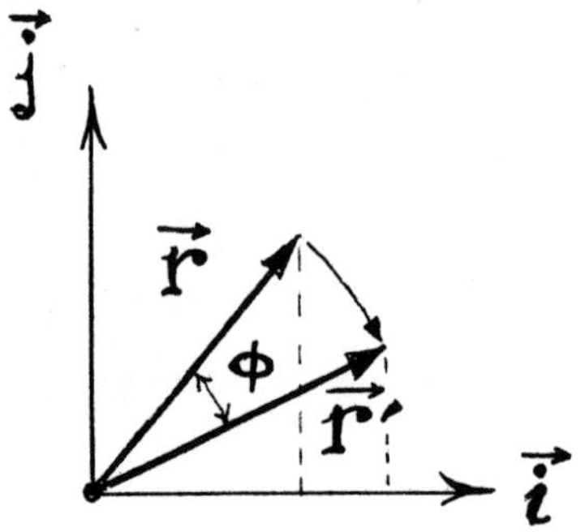

The operation S can also be regarded as a rotation of the **vector** in a **stationary coordinate system.**

From this point of view the operation

$$\mathbf{r}' = S \cdot \mathbf{r}$$

provides a **new vector r′** represented in terms of the **original bases.**

Thus S can be thought of as a clockwise rotation of the vector **r** through an angle ϕ.

With such an *interpretation* the base vectors associated with the final vector **r′** are the original bases **i** and **j.**

$$\begin{bmatrix} x_1' \\ x_2' \end{bmatrix} = \begin{pmatrix} S_{11} & S_{12} \\ S_{21} & S_{22} \end{pmatrix} \begin{bmatrix} x_1 \\ x_2 \end{bmatrix}$$

with

$$\mathbf{r} = x_1\mathbf{i} + x_2\mathbf{j}$$

and with

$$\mathbf{r}' = x_1'\mathbf{i} + x_2'\mathbf{j}$$

This possibility of a double interpretation leads to little difficulty in the end result if the questions and conditions of the problem are clearly stated.

CHAPTER 3

Analytic Geometry

A. Introduction

IN CHAPTER 1 WE mentioned that **analytic geometry** is an algebraic description of geometrical figures in E_2 and E_3. Some material which is considered a part of analytic geometry has been introduced in our discussion of Geometry and Matrices in Chapter 1.

We also find that a portion of traditional Coordinate Geometry has been anticipated in Chapter 2 which concerned the Vector Algebra.

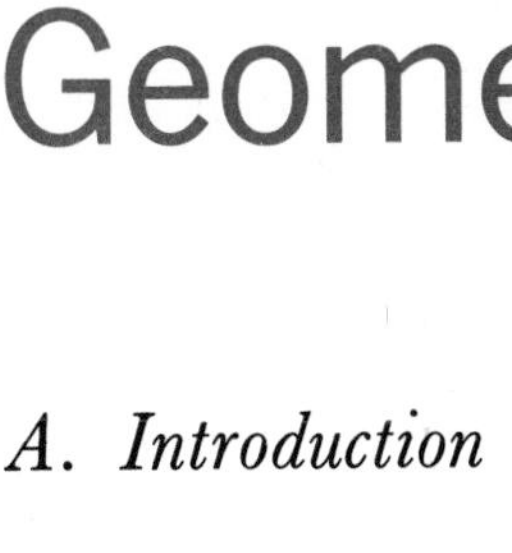

The term Analytic Geometry has come to imply the study of certain elementary figures in E_2 and E_3, namely, those of the **linear** and **quadratic** varieties.

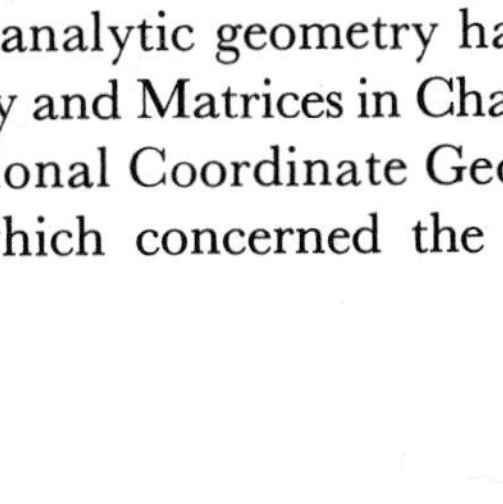

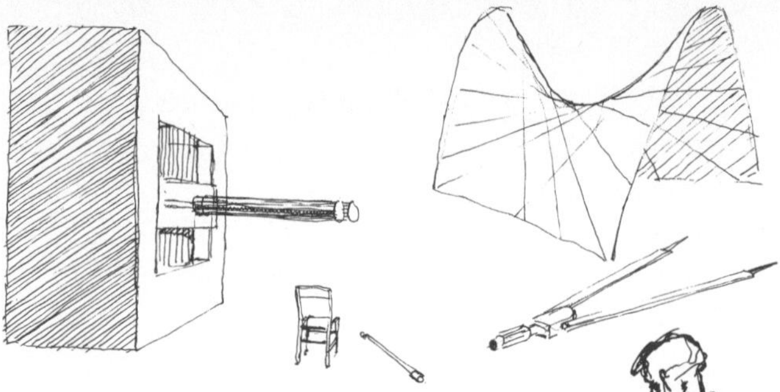

In E_2 the linear figures are *straight lines,* and the quadratic figures are the *conic sections* (ellipses, hyperbolas, parabolas, and circles).

In E_3 the linear equations represent the *planes,* while the quadratic figures are the ellipsoids, spheres, parabloids, etc.

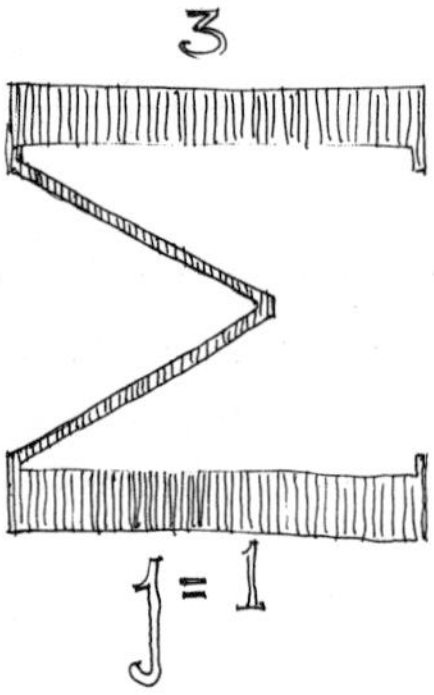

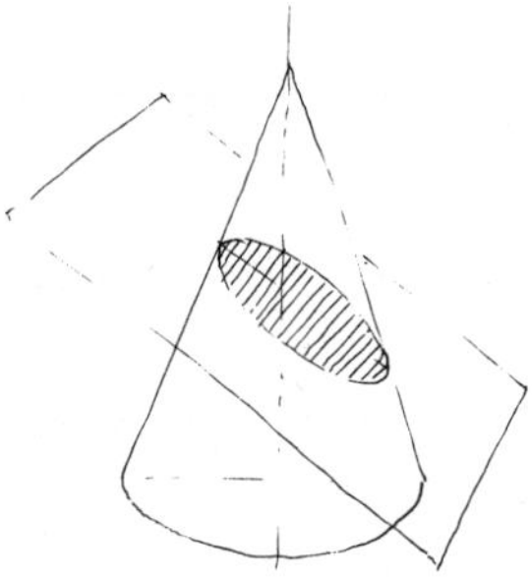

B. Loci

The term Locus in analytic geometry refers to the figure composed of **all** of the **points** satisfying certain conditions. These conditions can be specified as either geometrical or algebraic conditions.

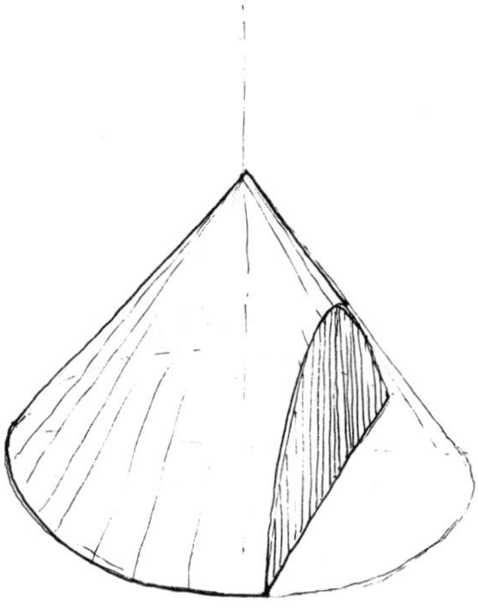

As an example of the former; the **locus** of all points in E_3 at a fixed perpendicular distance R from a given straight line is clearly a cylinder of radius R with the given line as the axis.

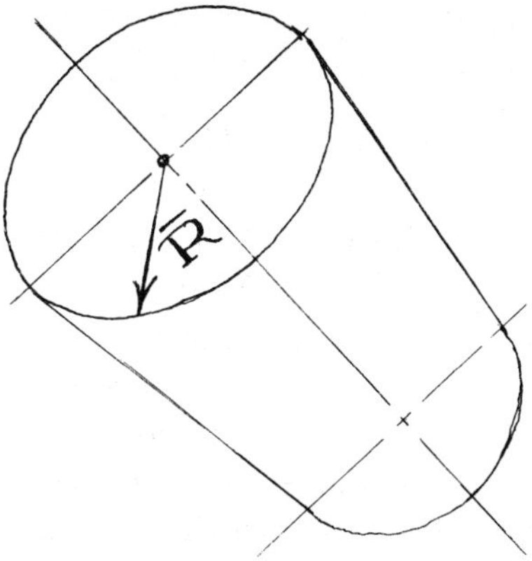

To illustrate the locus defined by algebraic equations; let x, y be a set of Cartesian coordinates in E_2, and let C be the **locus** of all of the points in E_2 whose coordinates satisfy the relation.

$$x^2 + y^2 = 1$$

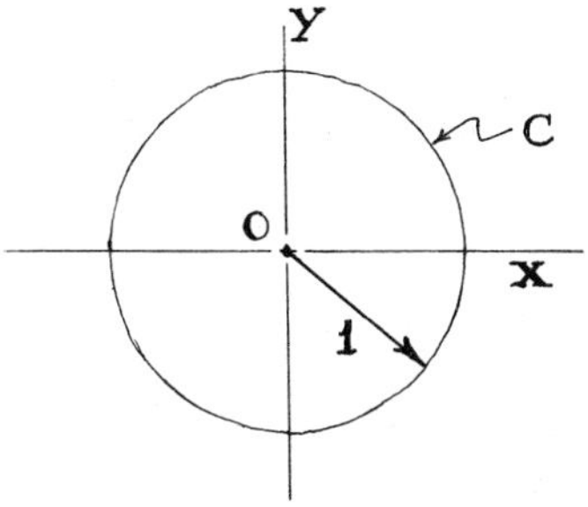

C is clearly a circle of unit radius with its center at the origin, for $\sqrt{x^2 + y^2}$ is simply the distance of the point (x, y) from the origin O.

Analytic geometry is concerned primarily with loci as specified by algebraic equations. This is by far the most useful technique. Shortly we shall see that loci specified by geometrical conditions can usually be specified equally well by algebraic relations.

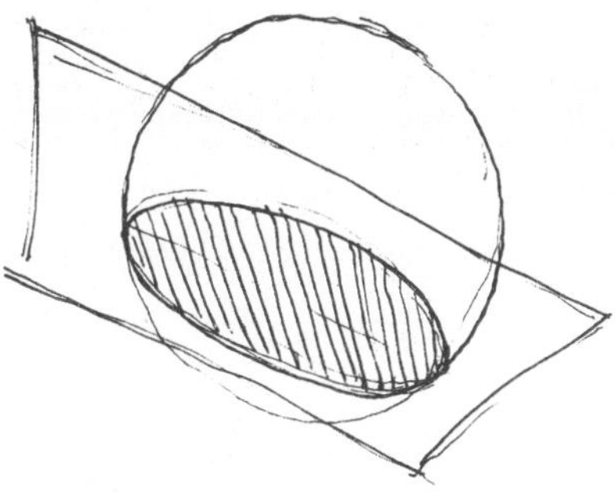

C. Straight Lines

Instead of discussing the properties of the equation for a straight line in E_2 and later extending the discussion to E_3 we will utilize the vector technique to write down the general form for a line in E_3. With this in view we can later as examples develop the various strictly algebraic forms.

There are several methods by which a straight line can be represented vectorially.

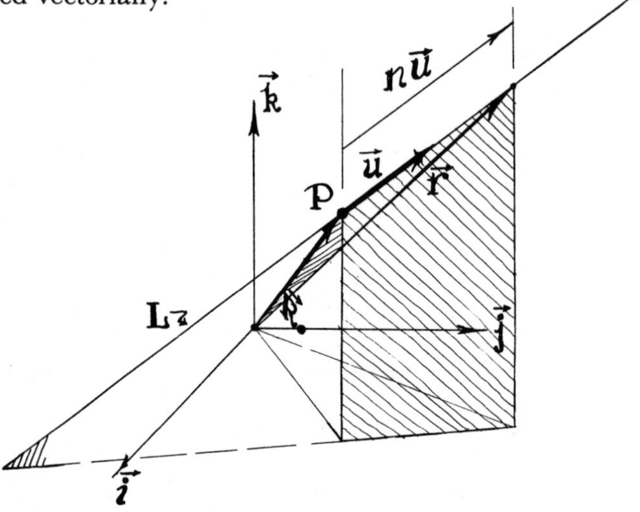

By and large the most useful of these representations is that particular one which uses the definition of the vector as a DIRECTED LINE SEGMENT to describe the straight line.

Consider the line L; the point P lies upon L, and the vector **u** originating at P is constructed to lie along L.

The general description of the line L is provided by an equation for the position vector **r** which **terminates only upon L.**

The definition of **r** can be obtained by a **vector addition** of the vector $\mathbf{R}_0$ which locates the point P and a vector n**u** lying upon L. Notice that n is a number (positive or negative) which varies the magnitude of the vector **u.**

In general therefore

$$\mathbf{r} = n\mathbf{u} + \mathbf{R}_0$$

Thus whenever we are given a point on the line and a vector parallel to the line we can define the line by the vector equation above.

Let

$$\mathbf{R}_o = x_o\mathbf{i} + y_o\mathbf{j} + z_o\mathbf{k}$$

and let

$$\mathbf{u} = \alpha\mathbf{i} + \beta\mathbf{j} + \gamma\mathbf{k}$$

then

$$\mathbf{r} = (n\alpha + x_o)\,\mathbf{i} + (n\beta + y_o)\,\mathbf{j} + (n\gamma + z_o)\,\mathbf{k}$$

An equivalent equation is

$$\mathbf{r} = x\mathbf{i} + y\mathbf{j} + z\mathbf{k}$$

Therefore we can write

$$x = n\alpha + x_o$$
$$y = n\beta + y_o$$
$$z = n\gamma + z_o$$

These are called the **parametric equations** for the line L. The variable parameter is n.

In the most general case α, β, and γ are numbers.

If we define the vector **u** as a **unit vector** such that

$$\alpha^2 + \beta^2 + \gamma^2 = 1$$

then n is the **distance** from P to the point of termination of **r**.

In other words under these conditions n is distance from (x_o, y_o, z_o) to (x, y, z).

Now consider two lines L and L′ intersecting at P

Let L be defined by

$$\mathbf{r} = n\mathbf{u} + \mathbf{R}_o$$

and let L′ be defined by

$$\mathbf{r} = m\mathbf{v} + \mathbf{R}_o$$

where

$$\mathbf{v} = \alpha'\mathbf{i} + \beta'\mathbf{j} + \gamma'\mathbf{k}$$

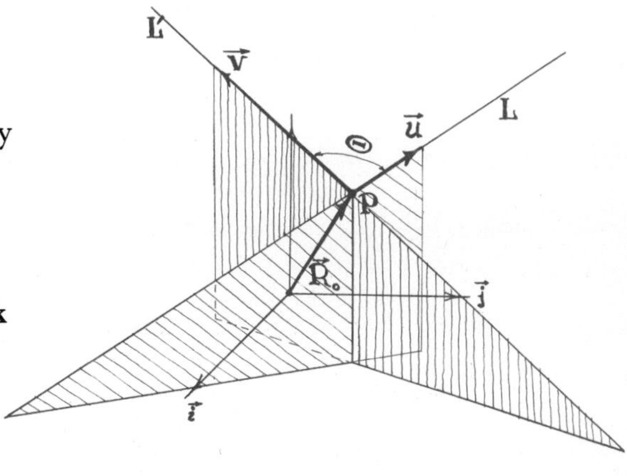

If both **u** and **v** are unit vectors the cosine of the angle θ between L and L′ is found quite readily by taking the **dot product** of **u** and **v.**

$$\mathbf{u} \cdot \mathbf{v} = |\mathbf{u}||\mathbf{v}| \cos\theta = \alpha\alpha' + \beta\beta' + \gamma\gamma'$$

If $|\mathbf{u}| = |\mathbf{v}| = 1$ then

$$\cos\theta = \alpha\alpha' + \beta\beta' + \gamma\gamma'$$

In case **u** and **v** have not been normalized (made into unit vectors) then

$$\cos\theta = \frac{\alpha\alpha' + \beta\beta' + \gamma\gamma'}{\sqrt{\alpha^2 + \beta^2 + \gamma^2}\ \sqrt{\alpha^{2'} + \beta^{2'} + \gamma^{2'}}}$$

Suppose that **u** and **v** are unit vectors. We can illustrate the power of this representation by finding the distance between a point Q_1 on L and a point Q_2 on L′, where Q_1 is located a distance n from P and Q_2 a distance m from P. By the law of cosines $d^2 = m^2 + n^2 - 2mn\cos\theta = m^2 + n^2 - 2mn(\alpha\alpha' + \beta\beta' + \gamma\gamma')$.

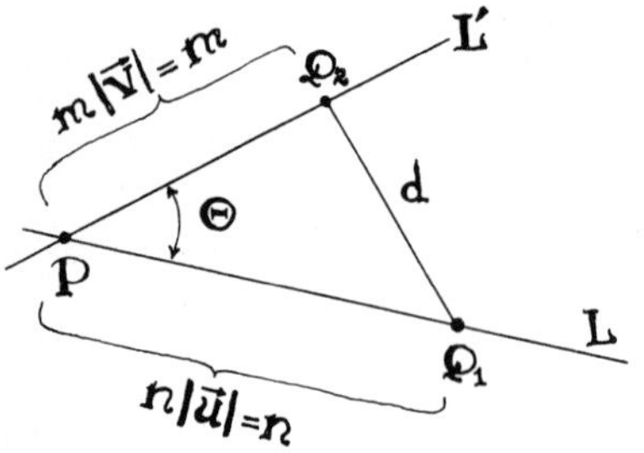

As a numerical example of these manipulations compute the angle θ when $P = (1, 2, 3)$, $Q_1 = (3, 0, 1)$, and $Q_2 = (0, -1, 2)$.

The solution is readily obtained if we first construct the parametric vector form for the line.

First consider L (defined by **U**)

$$\mathbf{R}_o = \mathbf{i} + 2\mathbf{j} + 3\mathbf{k}$$

The vector **U** connecting the two points P and Q_1 is

$$\mathbf{U} = (3 - 1)\,\mathbf{i} + (0 - 2)\,\mathbf{j} + (1 - 3)\,\mathbf{k} = 2\mathbf{i} - 2\mathbf{j} - 2\mathbf{k}$$

We can convert **U** to the unit vector **u**;

$$\mathbf{u} = \frac{\mathbf{U}}{|\mathbf{U}|} = \frac{2\mathbf{i} - 2\mathbf{j} - 2\mathbf{k}}{\sqrt{4 + 4 + 4}} = \frac{1}{3}\{\mathbf{i} - \mathbf{j} - \mathbf{k}\}$$

Then L is given by

$$\mathbf{r} = n\frac{1}{\sqrt{3}}\{\mathbf{i} - \mathbf{j} - \mathbf{k}\} + \{\mathbf{i} + 2\mathbf{j} + 3\mathbf{k}\}$$

or
$$x = \frac{n}{\sqrt{3}} + 1$$
$$y = \frac{-n}{\sqrt{3}} + 2$$
$$z = \frac{-n}{\sqrt{3}} + 3$$

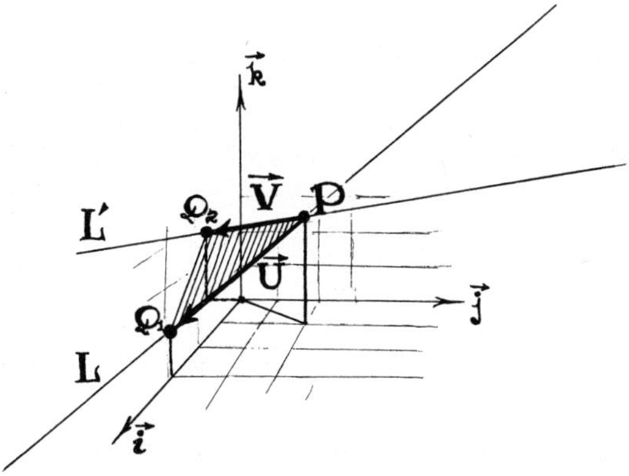

In the same manner we can define L′.

$$\mathbf{V} = (0 - 1)\,\mathbf{i} + (-1 - 2)\,\mathbf{j} + (2 - 3)\,\mathbf{k} = -\mathbf{i} - 3\mathbf{j} - \mathbf{k}$$

The unit vector **v** is

$$\mathbf{v} = \frac{\mathbf{V}}{|\mathbf{V}|} = \frac{-\mathbf{i} - 3\mathbf{j} - \mathbf{k}}{\sqrt{1 + 9 + 1}} = \frac{1}{\sqrt{11}}\{-\mathbf{i} - 3\mathbf{j} - \mathbf{k}\}$$

Thus L′ is given by

$$r = \frac{m}{\sqrt{11}}\{-1\mathbf{i} - 3\mathbf{j} - \mathbf{k}\} + \{\mathbf{i} + 2\mathbf{j} + 3\mathbf{k}\}$$

or
$$x = \frac{-m}{\sqrt{11}} + 1$$
$$y = \frac{-3m}{\sqrt{11}} + 2$$
$$z = \frac{-m}{\sqrt{11}} + 3$$

The cosine of the angle between L and L′ is merely

$$\mathbf{u} \cdot \mathbf{v} = \cos\phi = \frac{(\mathbf{i} - \mathbf{j} - \mathbf{k})}{\sqrt{3}} \times \frac{(-\mathbf{i} - 3\mathbf{j} - \mathbf{k})}{\sqrt{11}}$$

or

$$\cos\phi = \frac{(-1 + 3 + 1)}{\sqrt{33}} = \frac{3}{\sqrt{33}}$$

We have shown how one constructs a line between two points P and Q in the previous example. It is important to recognize that this method of analysis (the parametric vector) is the only method which holds equally well in E_3 or E_2. Other representations which we shall consider must carry conditions limiting their application to one space or the other.

Because the slope of the line in two dimensions is of utmost importance we can now examine our general parametric form in two dimensions so that we can see most clearly the role of the line slope.

Regard the point at which L intersects the x axis and the angle ϕ which L makes with the x axis.

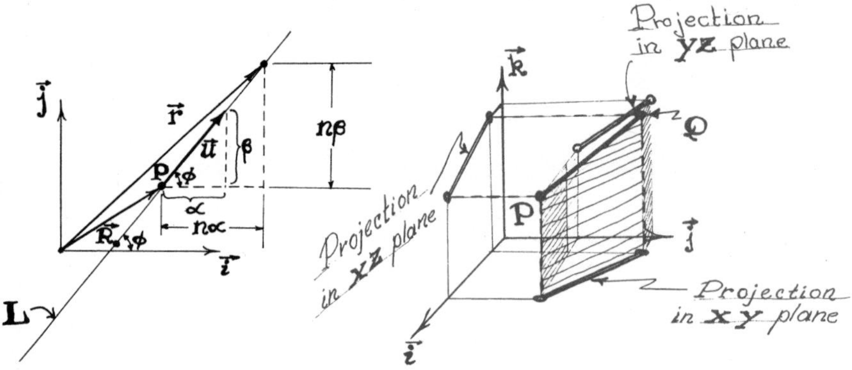

The slope of L in two dimensions is defined as the tangent of ϕ. Using the **vector parametric form** with **u** a unit vector we see that the slope of L is given by

$$\tan\phi = \text{slope} = \frac{\mathbf{u} \cdot \mathbf{j}}{\mathbf{u} \cdot \mathbf{i}} = \frac{\beta}{\alpha}$$

In three dimensions we can speak of the slope of the projections of L upon the 3 possible planes; (x,y), (y,z), and (z,x).

In two dimensions we see that if the slope ($\tan \phi$) is given then **u** the unit vector is specified.

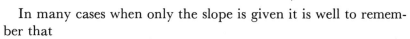

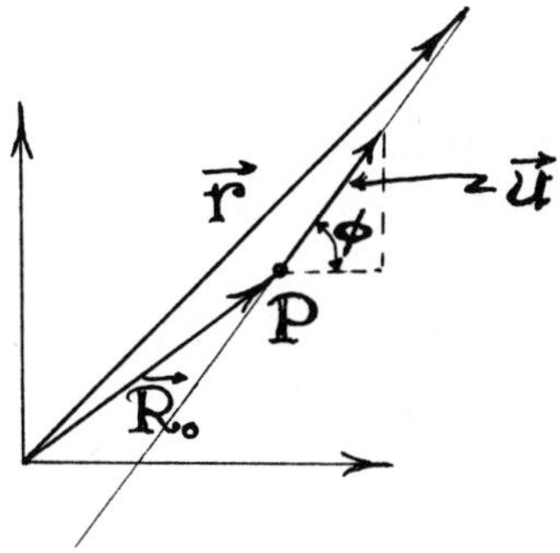

Since $\tan\phi = \beta/\alpha$

then $\cos\phi = \alpha$

and $\sin\phi = \beta$

We then can write

$$\mathbf{u} = \cos\phi\, \mathbf{i} + \sin\phi\, \mathbf{j}$$

In many cases when only the slope is given it is well to remember that

$$\cos \phi = \frac{1}{\sqrt{1 + \tan^2\phi}}$$

$$\sin \phi = \frac{\tan\phi}{\sqrt{1 + \tan^2\phi}}$$

thus

$$\mathbf{r} = \frac{1}{\sqrt{1 + \tan^2\phi}} \{\mathbf{i} + \tan\phi\, \mathbf{j}\} + \mathbf{R}_o$$

The customary equation that is utilized to represent a line L in two dimensions is the linear algebraic equation in two unknowns.

The locus of all points in E_2 whose coordinates satisfy an algebraic equation of the type

$$ax + by + c = O$$

is a straight line, where a, b, and c are numbers with a and b not simultaneously zero (otherwise our equation reduces to $c = o$).

The parametric equations obtained from the vector form are,

$$x = n\alpha + x_o$$
$$y = n\beta + y_o$$

If the parameter n is eliminated between these two equations we find that

$$n = \frac{y - y_0}{\beta} = \frac{(x - x_0)}{\alpha}$$

and $$y - y_0 = \frac{\beta}{\alpha}(x - x_0) = \tan \phi \, (x - x_0)$$

This equation has been named the point-slope equation and as we see it arises naturally from our more general form. By further manipulation we obtain the linear form

$$\beta x - \alpha y - \beta x_0 + \alpha y_0 = O$$

Thus β is proportional to a

$-\alpha$ is proportional to b

and $-\beta x_0 + \alpha y_0$ is proportional to c

In three dimensions the elimination of the parameter n serves to give three algebraic relations.

$$n = \frac{(x - x_0)}{\alpha} = \frac{(y - y_0)}{\beta} = \frac{(z - z_0)}{\gamma}$$

Later we will see that these equations result from the two simultaneous linear algebraic equations in **three** unknowns. In three dimensions the linear equation represents a **plane.** Thus two simultaneous linear equations represent the **intersection of two planes,** which is a straight line.

Representation of a line in E_2 by the perpendicular vector

A more restricted vector form for a straight line in two dimensions can be set up using the linear algebraic form directly.

Construct a constant vector

$$\mathbf{W} = a\,\mathbf{i} + b\,\mathbf{j}$$

If all of the points (x, y) are represented as a position vector

$$\mathbf{r} = x\,\mathbf{i} + y\,\mathbf{j}$$

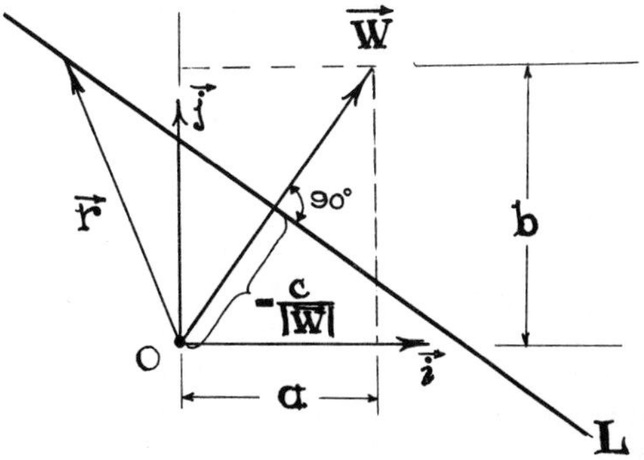

The equation of the straight line is written

$$\mathbf{W} \cdot \mathbf{r} = -c$$

At first glance we see that $\mathbf{W} \cdot \mathbf{r} = ax + by = -c$ which agrees with our original linear form.

The vector equation has several interesting properties.

1. The line defined by $\mathbf{W} \cdot \mathbf{r} = -c$ consists of the terminal points of *all* of the vectors **r**.
2. The line is perpendicular to **W.**
3. The line intersects **W** at a distance $\frac{-c}{|\mathbf{W}|}$ from the base point O.

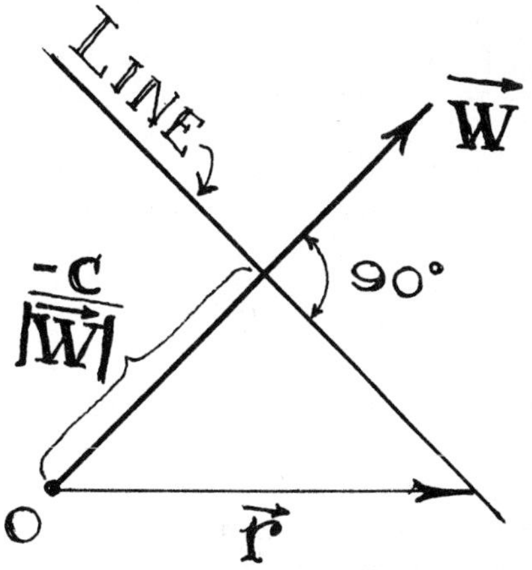

An important manipulation which one must be prepared to make is the transformation from one characteristic form to another. This is readily achieved if we recall that

"a" is proportional to β

and

b is proportional to $-\ \alpha$

Since the components of **W** are a and b we can construct the components of **u** in the parametric form in the following way:

$$\alpha = \frac{-\ b}{\sqrt{a^2 + b^2}}$$

and

$$\beta = \frac{a}{\sqrt{a^2 + b^2}}$$

Since $c = -\beta x_0 + \alpha y_0$, knowing α and β we can find $\mathbf{R}_0$ by choosing a particular x_0 and solving for y_0;

$$y_0 = \frac{c + \beta x_0}{\alpha}$$

To illustrate the use of the method of the perpendicular vector take the following problem:

Find the distance from the origin to the line.

$$2x + 3y - 6 = O$$

The perpendicular vector from O is

$$W = 2\,\mathbf{i} + 3\,\mathbf{j}$$

The distance from O to L is the value of the projection of $\mathbf{r}$ upon $\mathbf{W}$.

In other words we want $\dfrac{\mathbf{r} \cdot \mathbf{W}}{|\mathbf{W}|}$.

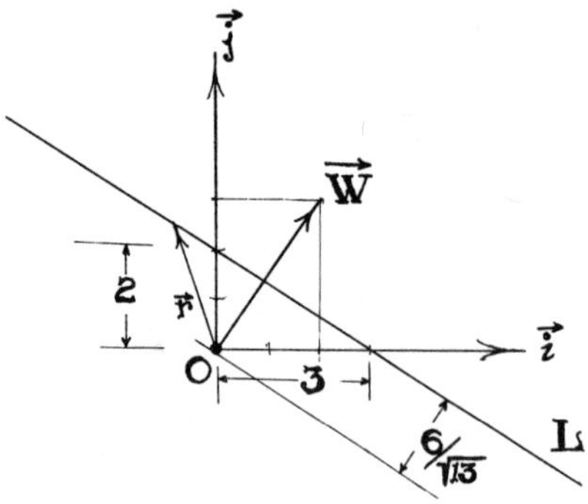

But this is merely $\dfrac{-c}{|\mathbf{W}|}$, or

$$\frac{6}{\sqrt{4+9}} = \frac{6}{\sqrt{13}}$$

D. The Plane

The plane in E_3 is the locus of all points satisfying the relation

$$ax + by + cz + d = O$$

assuming that a, b, and c are not simultaneously zero.

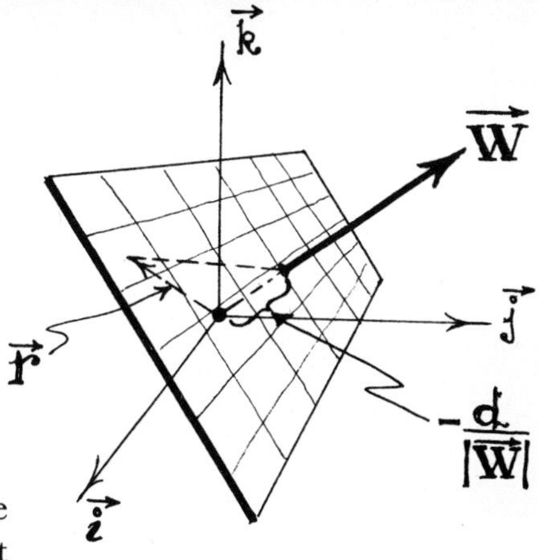

The vector equation

$$\mathbf{W} \cdot \mathbf{r} + d = O$$

where

$$\mathbf{W} = a\,\mathbf{i} + b\,\mathbf{j} + c\,\mathbf{k}$$

and

$$\mathbf{r} = x\,\mathbf{i} + y\,\mathbf{j} + z\,\mathbf{k}$$

is the general equation defining the plane in three dimensions. If we set $z = O$ this form becomes the equation for the straight line in the xy plane, since this condition of $z = O$ defines the intersection of the plane $\mathbf{W} \cdot \mathbf{r} + d = O$ and the xy plane.

The perpendicular distance of the plane from the origin is given by the projection of **r** on **W**,

$$\perp \text{ distance* } O \rightarrow \text{plane} = \mathbf{r} \cdot \frac{\mathbf{W}}{|\mathbf{W}|} = -\frac{d}{|\mathbf{W}|}$$

It is clear in all of these cases that if **W** is constructed as a unit vector **w**,

$$\mathbf{w} = \frac{\mathbf{W}}{|\mathbf{W}|}$$

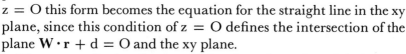

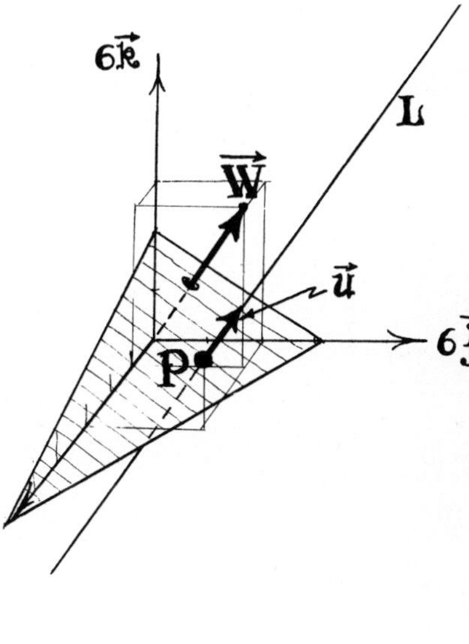

Then in the equation

$$\mathbf{w} \cdot \mathbf{r} + \delta = O$$

the magnitude of δ always gives the perpendicular distance in question.

As an exercise obtain the equation of the line through the point (3, 2, 1) which is perpendicular to the plane.

$$x + 2y + 3z - 6 = O$$

* The sign $\perp$ is shorthand for the word **perpendicular.**

Notice first that **W** is perpendicular to the plane;

$$\mathbf{W} = \mathbf{i} + 2\mathbf{j} + 3\mathbf{k}$$

We normalize; and the vector **u** of the line L is parallel to the perpendicular vector **w.**

$$\mathbf{u} = \mathbf{w} = \frac{\mathbf{i} + 2\mathbf{j} + 3\mathbf{k}}{\sqrt{1+4+9}} = \frac{1}{\sqrt{14}}(\mathbf{i} + 2\mathbf{j} + 3\mathbf{k})$$

Because L must pass through (3, 2, 1) we know $\mathbf{R}_o$,

$$\mathbf{R}_o = 3\mathbf{i} + 2\mathbf{j} + \mathbf{k}$$

Therefore the parametric equation of L is

$$\mathbf{r} = n\mathbf{u} + \mathbf{R}_o = \left(\frac{n}{\sqrt{14}} + 3\right)\mathbf{i} + \left(\frac{2n}{\sqrt{14}} + 2\right)\mathbf{j} + \left(\frac{3n}{\sqrt{14}} + 1\right)\mathbf{k}$$

Then

$$n = \sqrt{14}\,(x-3) = \frac{\sqrt{14}}{2}(y-2) = \frac{\sqrt{14}}{3}(z-1)$$

or

$$(x-3) = \frac{(y-2)}{2} = \frac{(z-1)}{3}$$

E. Curves in E_2

Conic Sections

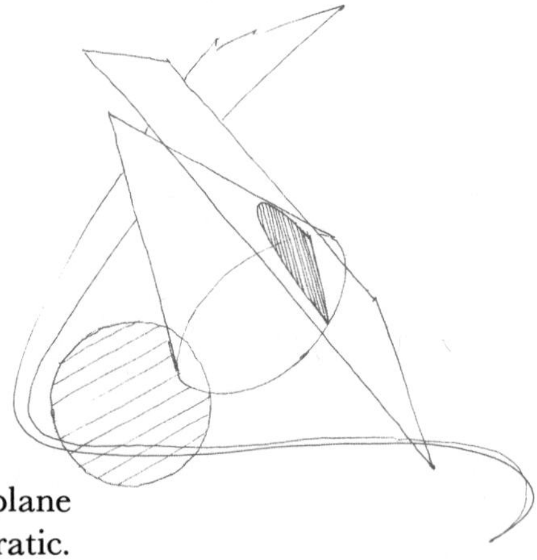

Historically the general field of quadratic equations in two dimensions was given the title of conic sections because the curves formed by the intersection of a cone and a plane in E_3 define a two dimensional quadratic.

In addition to these pictorial characteristics of the two dimensional quadratic the curves belonging to this group are customarily defined in terms of length invariants, such as the **ellipse** which is generally defined as the LOCUS consisting of all points Q such that the sum of the distances between Q and **two fixed points** P_1 and P_2 is a constant.

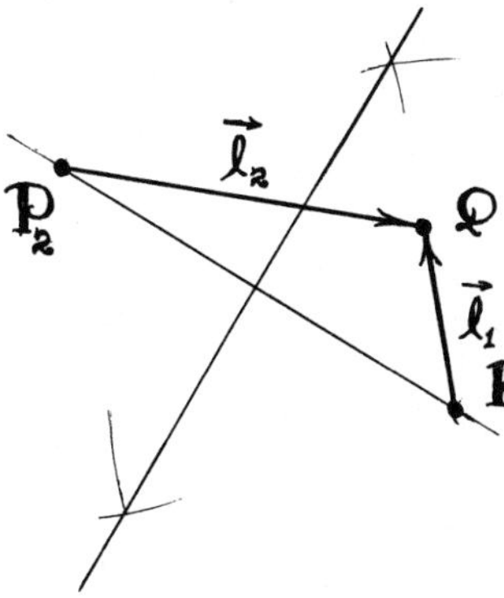

From such approaches one obtains very special quadratic forms which later can be generalized.

The Ellipse

A brief mention will be made of the **locus** definition of the Ellipse.

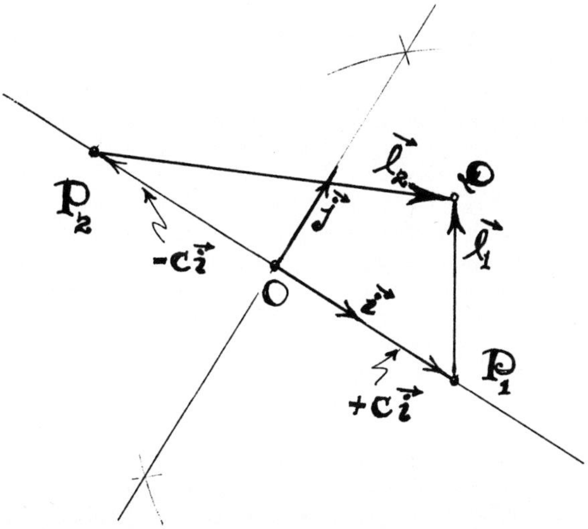

The Ellipse is the **locus** consisting of all points Q such that the distance $|\mathbf{l}_1|$ of Q to P_1 **plus** the distance $|\mathbf{l}_2|$ of Q to P_2 is a fixed number K (where $K > 2c$; 2c is the distance between P_1 and P_2).

The origin of coordinates O has been placed at the point of symmetry, and P_1 and P_2 have been constructed upon the x_1 axis at distances $\pm$ c from O.

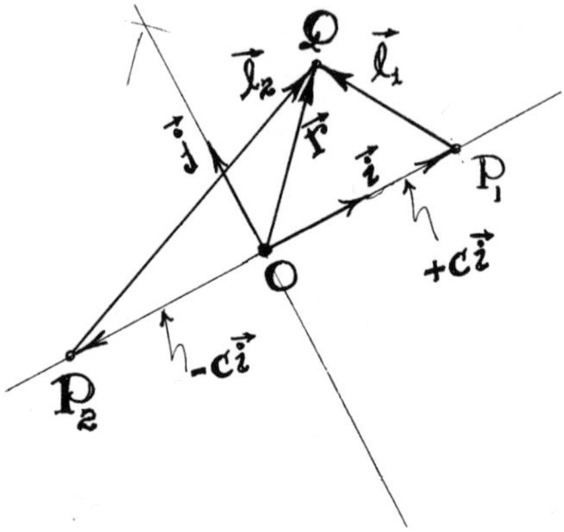

Because

$$\mathbf{l}_1 = \mathbf{r} - c\mathbf{i}$$

and

$$\mathbf{l}_2 = \mathbf{r} + c\mathbf{i}$$

where

$$\mathbf{r} = x_1\mathbf{i} + x_2\mathbf{j}$$

the constraint

$$|\mathbf{l}_1| + |\mathbf{l}_2| = K$$

becomes

$$|\mathbf{r} - c\,\mathbf{i}| + |\mathbf{r} + c\,\mathbf{i}| = K$$

or

$$\sqrt{(x_1 - c)^2 + x_2{}^2} + \sqrt{(x_1 + c)^2 + x_2{}^2} = K$$

After squaring both sides of this equation we rearrange to maintain the square root on one side, and then square again. The result is

$$\frac{x_1{}^2}{a^2} + \frac{x_2{}^2}{b^2} = 1$$

where

$$a = K/2 \text{ and } b = \sqrt{a^2 - c^2}$$

This form is called **the normal form** for the ellipse. This particular form is a special case of the two dimensional quadratic with $A_{11} = 1/(a)^2$, $A_{12} = A_{21} = O$, $A_{22} = 1/(b)^2$.

It is easy to perceive that the ellipse crosses the x axis at (a, o) and (−a, o) while it crosses the y axis at (o, b) and (o, −b). The Points (c, o) and (−c, o) are called the **foci** of the **ellipse.**

a = the length of the **semi-major** Axis
b = the length of the **semi-minor** Axis.

We observe that b would be imaginary if $a^2 < c^2$, or $K < 2c$.

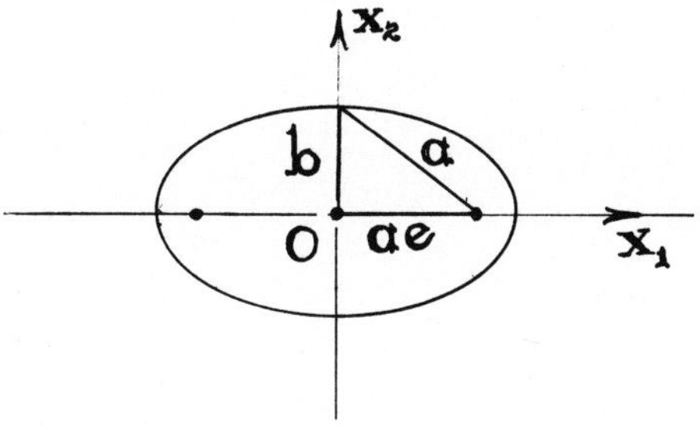

The **eccentricity of the ellipse** is denoted by e and defined as

$$e = \frac{c}{a} = \frac{\sqrt{a^2 - b^2}}{a} \quad \left\{ \begin{array}{l} \text{for } b < a \text{, focii on the} \\ \text{x axis.} \end{array} \right\}$$

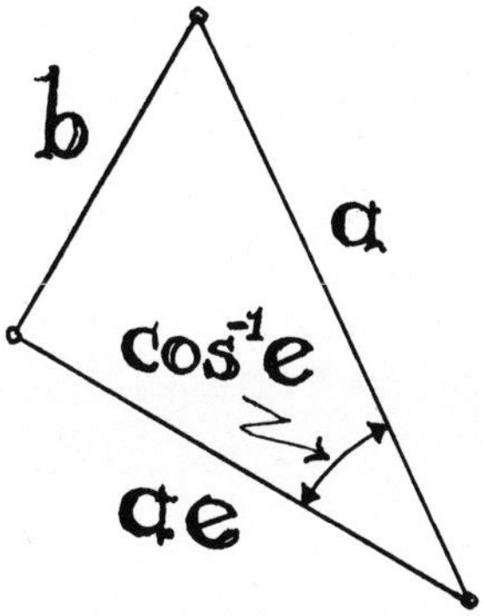

a is the hypotenuse of a right triangle having legs b and ae

$$O \leq e < 1 \text{ for the ellipse.}$$

The circle is a degenerate form of the ellipse; it arises when

$$e = o \text{ and } b = a$$

The Hyperbola

In the previous section we asked for the locus of all points the *sum* of whose distances from two points was a constant K. To obtain the **hyperbola** we ask a similar question.

What is the locus of all points Q such that the **difference** of the **distances** of Q to **two** fixed points P_1 and P_2 is a **constant?**

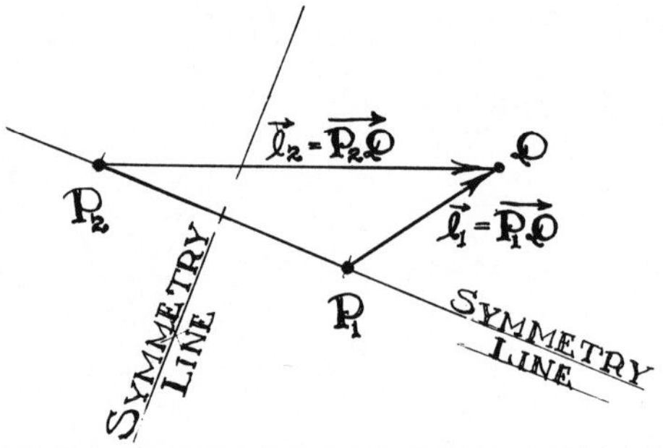

Again because of the symmetry set the two points P_1 and P_2 on the x_1 axis at $x_1 = c$ and at $x_1 = -c$.

Then

$$\mathbf{l}_1 = (x_1 - c)\mathbf{i} + x_2\mathbf{j}$$
$$\mathbf{l}_2 = (x_1 + c)\mathbf{i} + x_2\mathbf{j}$$

with

$$|\mathbf{l}_1| - |\mathbf{l}_2| = \pm K$$

Once again we square rearrange and square again giving

$$\frac{x_1^2}{a^2} - \frac{x_2^2}{b^2} = 1$$

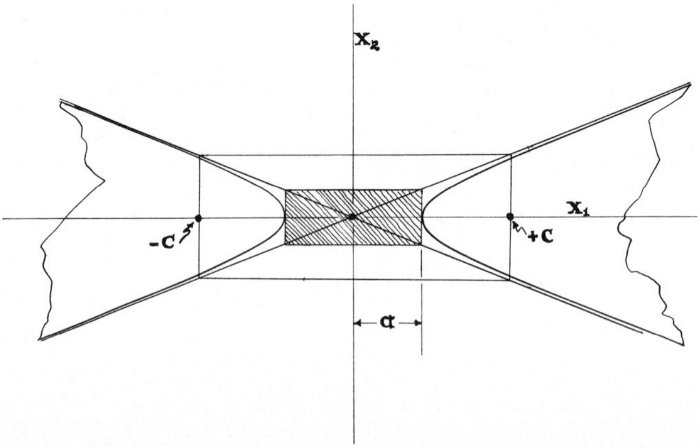

The **eccentricity** e is defined as in the case of the ellipse, noting that $b^2 \to -b^2$ thus

$$e = \frac{c}{a} = \frac{\sqrt{a^2 + b^2}}{a}$$

Pictorially we observe that the curves tend to **two straight lines** for **large values** of x_1 and x_2.

When $\frac{x_1}{a}$ and $\frac{x_2}{b}$ are much larger than 1 the equation for the hyperbola can be approximated by,

$$\frac{x_1^2}{a^2} - \frac{x_2^2}{b^2} \simeq 0$$

Thus for large distances from the origin

$$x_2 \simeq \pm \frac{b}{a} x_1$$

defining the lines toward which the curves tend. These lines are called **asymptotes.**

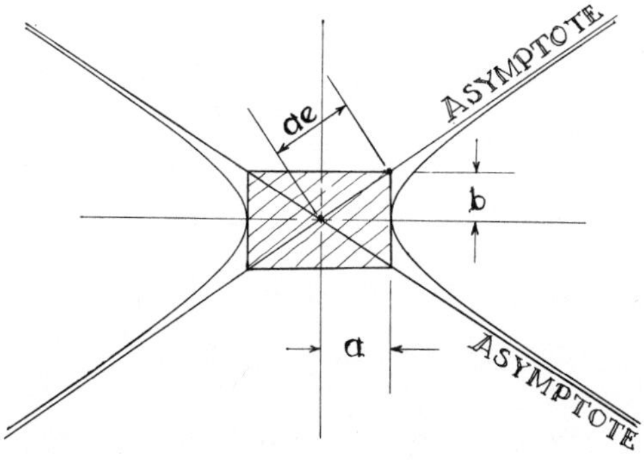

If we change the signs such that

$$\frac{x_2^2}{b^2} - \frac{x_1^2}{a^2} = 1$$

we define a new set of hyperbolas with focii on the x_2 axis. This **set** is called the **conjugate set** to the first set which had focii on the x_1 axis.

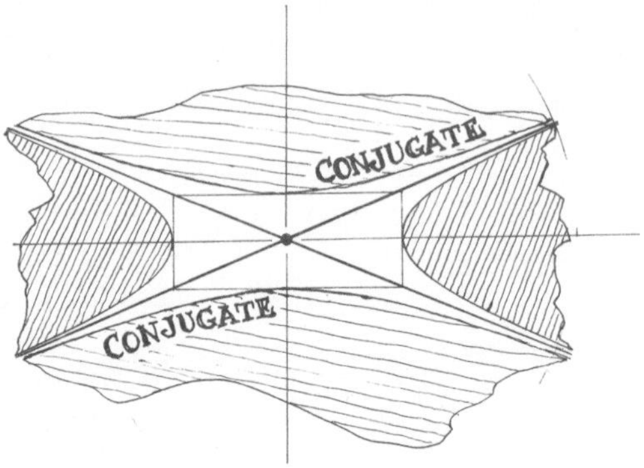

The Parabola

In the preceding cases we dealt with loci which were defined in terms of distances from two fixed points.

The Parabola is the locus of all points Q which are equi-distant from a point P and a line L.

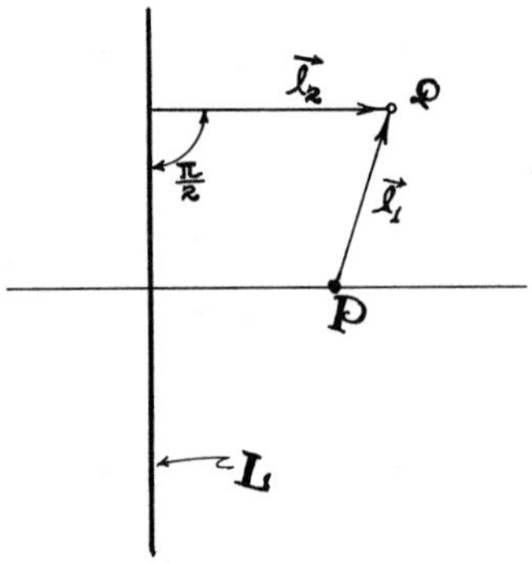

The only axis of symmetry in this problem is the line through P **perpendicular** to L. Therefore set this line along the x_1 axis while L is aligned with the x_2 axis (or y axis).

$$|\mathbf{l}| = l = |\mathbf{l}_1| = |\mathbf{l}_2|$$

Now

$$\mathbf{l}_1 = (x_1 - 2c)\,\mathbf{i} + x_2\,\mathbf{j}$$
$$\mathbf{l}_2 = x_1\,\mathbf{i}$$

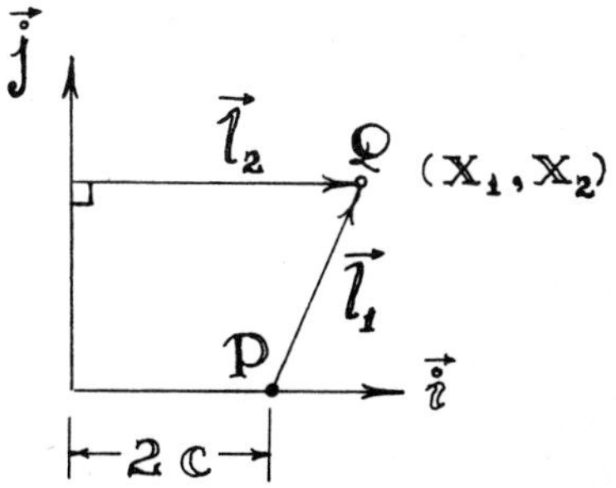

Thus

$$|\mathbf{l}_1| = |\mathbf{l}_2|$$

becomes

$$\sqrt{(x_1 - 2c)^2 + x_2^2} = \sqrt{x_1^2}$$

and
$$-4cx_1 + 4c^2 + x_2^2 = 0$$

Solving for x_2^2

$$x_2^2 = 4c\,\{x_1 - c\}$$

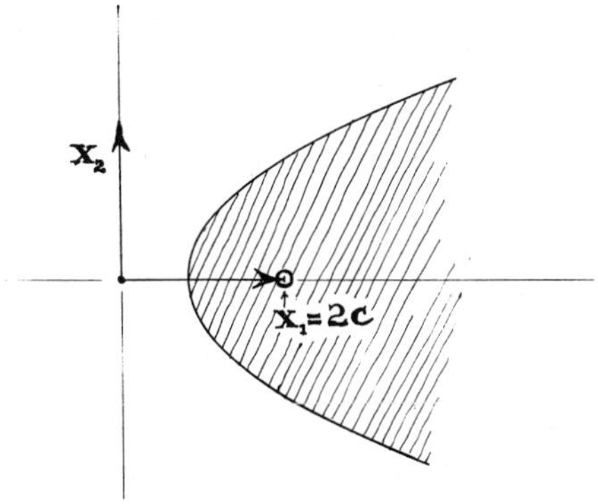

F. The Quadratic Form

Introduction

The reader should be cautioned because the strictly two dimensional geometric analysis in no way prepares him to anticipate the elegance and generality of the quadratic form in three or N dimensions. The world of science contains many subjects which are described in terms of quadratic forms, and the analysis of many of these subjects depends upon the utilization of the most general properties of the forms.

Inner products between vectors may result in quadratics of particular interest. A trivial example is the inner product of a vector with itself giving the square of the magnitude of that vector; for example in two dimensions using matrix representation

$$\tilde{\mathbf{r}} \cdot \mathbf{r} = [x_1, x_2] \begin{bmatrix} x_1 \\ x_2 \end{bmatrix} = x_1^2 + x_2^2$$

If a constraint is placed upon this operation which requires that the inner product be equal to a constant, the resulting equation is that of **the circle.**

$$\tilde{\mathbf{r}} \cdot \mathbf{r} = R^2 \text{ gives } x_1^2 + x_2^2 = R^2$$

In the example above we can think of the operation as one involving a transformation matrix. In this case the **identity** or **unit matrix** must be used.

$$\tilde{\mathbf{r}} \cdot \mathbf{r} = \tilde{\mathbf{r}} \cdot \mathbb{I} \cdot \mathbf{r} = [x_1, x_2] \begin{pmatrix} 1 & 0 \\ 0 & 1 \end{pmatrix} \begin{bmatrix} x_1 \\ x_2 \end{bmatrix} = x_1^2 + x_2^2 = R^2$$

On the other hand a more general two dimensional quadratic can be constructed by using a general matrix A instead of the unit matrix.

$$\begin{aligned} \tilde{\mathbf{r}} \cdot \mathbb{A} \cdot \mathbf{r} &= [x_1, x_2] \begin{pmatrix} A_{11} & A_{12} \\ A_{21} & A_{22} \end{pmatrix} \begin{bmatrix} x_1 \\ x_2 \end{bmatrix} \\ &= x_1 A_{11} x_1 + x_1 A_{12} x_2 + x_2 A_{21} x_1 + x_2 A_{22} x_2 \\ &= A_{11} x_1^2 + (A_{12} + A_{21})\, x_1\, x_2 + A_{22}\, x_2^2 \end{aligned}$$

The constraint that this equation be equal to a fixed number K provides a general two dimensional quadratic which is symmetric with respect to the origin.

Because we are dealing with **real** numbers (i.e., the values of x_1

and x_2 are real) and because one generally starts from the scalar quadratic equation no generality is lost by utilizing a **symmetric matrix** A,

$$A_{12} = A_{21}$$
$$\text{or} \qquad A_{mn} = A_{nm}$$

The most general equation for a conic section may contain linear terms describing the translation of the center of symmetry. The operation to be described pertains only to the ellipse and hyperbola. The parabola does *not* contain a *symmetry point;* thus the operation of translation to be described will not apply in this one case.

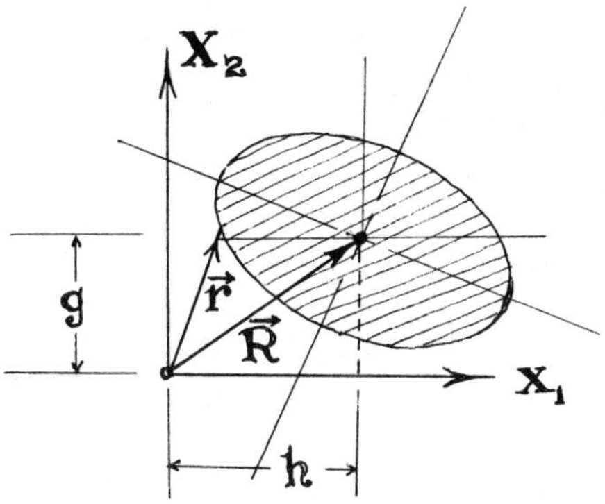

The **translated form** (using a symmetric A) is

$$A_{11}x_1^2 + 2A_{12}x_1x_2 + A_{22}x_2^2 + dx_1 + ex_2 + f = O$$

We first write this equation in the matrix form

$$\tilde{\mathbf{r}} \cdot \mathrm{A} \cdot \mathbf{r} + \tilde{\xi} \cdot \mathbf{r} + \mathrm{f} = \mathrm{O}$$

where ξ (a constant vector) is defined as

$$\xi = \begin{bmatrix} d \\ e \end{bmatrix}, \text{ or } \tilde{\xi} = [d, e]$$

Writing out the matrix form in detail,

$$[x_1, x_2]\begin{pmatrix} A_{11} & A_{12} \\ A_{21} & A_{22} \end{pmatrix}\begin{bmatrix} x_1 \\ x_2 \end{bmatrix} + [d, e]\begin{bmatrix} x_1 \\ x_2 \end{bmatrix} + f = O$$

The reader should keep in mind that

$$A_{12} = A_{21}$$

We assume further that the position of the symmetry point O′ is located by the position vector **R** having components h and g, i.e.,

$$\mathbf{R} = \begin{bmatrix} h \\ g \end{bmatrix}$$

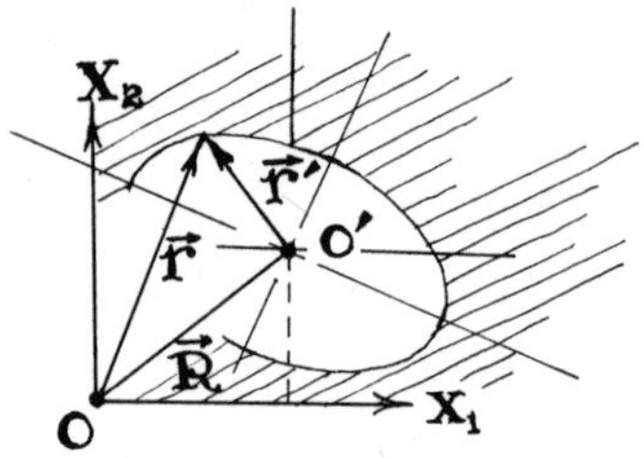

We can then replace **r** by the vector addition

$$\mathbf{r} = (\mathbf{R} + \mathbf{r}')^*$$

where

$$\mathbf{r}' \text{ has the variables } x_1' \text{ and } x_2'$$

Substituting into $\tilde{\mathbf{r}} \cdot \mathbb{A} \cdot \mathbf{r} + \tilde{\xi} \cdot \mathbf{r} + f = O$, we find

$$(\tilde{\mathbf{R}} + \tilde{\mathbf{r}}') \cdot \mathbb{A} \cdot (\mathbf{R} + \mathbf{r}') + \tilde{\xi} \cdot (\mathbf{R} + \mathbf{r}') + f$$

Expanding this equation,

$$\tilde{\mathbf{R}} \cdot \mathbb{A} \cdot \mathbf{R} + \tilde{\mathbf{R}} \cdot \mathbb{A} \cdot \mathbf{r}' + \tilde{\mathbf{r}}' \cdot \mathbb{A} \cdot \mathbf{R} + \tilde{\mathbf{r}}' \cdot \mathbb{A} \cdot \mathbf{r}' + \tilde{\xi} \cdot \mathbf{R} + \tilde{\xi} \cdot \mathbf{r}' + f = O$$

* Remember the rules for vector addition. If $\mathbf{R} = \begin{bmatrix} h \\ g \end{bmatrix}$ and $\mathbf{r}' = \begin{bmatrix} x_1' \\ x_2' \end{bmatrix}$ then $\mathbf{R} + \mathbf{r}' = \begin{bmatrix} h + x_1' \\ g + x_2' \end{bmatrix}$; or in the base vector representation $\mathbf{R} + \mathbf{r}' = (h + x_1')\,\mathbf{i}$ $(g + x_2')\,\mathbf{j}$.

The form of the quadratic when expanded about the symmetry point is

$$\tilde{\mathbf{r}}' \cdot \mathrm{A} \cdot \mathbf{r}' = \mathrm{K}$$

where K is a scalar.

If the expansion above is to reduce to this form the terms linear in $\mathbf{r}'$ must vanish and the constants must be grouped.

Thus

$$(\tilde{\mathbf{R}} \cdot \mathrm{A} + \tilde{\xi}) \cdot \mathbf{r}' + \tilde{\mathbf{r}}' \cdot \mathrm{A} \cdot \mathbf{R} = \mathrm{O}$$

and

$$\mathrm{f} + \tilde{\mathbf{R}} \cdot \mathrm{A} \cdot \mathbf{R} + \tilde{\xi} \cdot \mathbf{R} = -\mathrm{K}$$

Warning. ***At this point the reader must exercise his knowledge of matrices. The results of these two equations will be developed to give the components of*** R ***and the constant*** K, ***but all of the rules of matrix multiplication and manipulation as outlined in Chapter I must be used.***

Consider the first equation,

$$(\tilde{\mathbf{R}} \cdot \mathrm{A} + \tilde{\xi}) \cdot \mathbf{r}' + \tilde{\mathbf{r}}' \cdot (\mathrm{A} \cdot \mathbf{R}) = \mathrm{O}$$

We first note that $\tilde{\mathbf{R}} \cdot \mathrm{A} = \widetilde{(\mathrm{A} \cdot \mathbf{R})}$, then

$$(\widetilde{(\mathrm{A} \cdot \mathbf{R})} + \tilde{\xi}) \cdot \mathbf{r}' + \tilde{\mathbf{r}}' \cdot (\mathrm{A} \cdot \mathbf{R}) = \mathrm{O}$$

$\mathrm{A} \cdot \mathbf{R}$ is a vector say **p.** Since the variables of $\mathbf{r}'$ are real we note that

$$\tilde{\mathbf{r}}' \cdot \mathbf{p} = \tilde{\mathbf{p}} \cdot \mathbf{r}'$$

or

$$\tilde{\mathbf{r}}' \cdot (\mathrm{A} \cdot \mathbf{R}) = \widetilde{(\mathrm{A} \cdot \mathbf{R})} \cdot \mathbf{r}'$$

Thus our equation simplifies to

$$\{2\,\widetilde{(\mathrm{A} \cdot \mathbf{R})} + \tilde{\xi}\} \cdot \mathbf{r}' = \mathrm{O}$$

Because the components of $\mathbf{r}'$ are arbitrary the vector in curly brackets must be zero.

$$2\,(\widetilde{\mathrm{A}\cdot\mathrm{R}}) + \tilde{\xi} = \mathrm{O}$$

or

$$\mathrm{A}\cdot\mathbf{R} = -\frac{1}{2}\xi$$

Under the assumption that A is *not* a zero matrix, the last equation can be cleared in order to obtain the components of **R**, the translation vector.

$$\mathrm{A}^{-1}\cdot\mathrm{A}\cdot\mathbf{R} = \mathbf{R} = -\frac{1}{2}\mathrm{A}^{-1}\cdot\xi$$

In Chapter I the components of the inverse matrix A^{-1} were derived

$$\mathrm{A}_{kn}' = \frac{(-1)^{k+n}\,\text{minor}\ \mathrm{A}_{nk}}{\text{Det}\ \mathrm{A}}$$

Since

$$\mathrm{A} = \begin{pmatrix} \mathrm{A}_{11} & \mathrm{A}_{12} \\ \mathrm{A}_{21} & \mathrm{A}_{22} \end{pmatrix}$$

$$\mathrm{A}^{-1} = \frac{\begin{pmatrix} \mathrm{A}_{22} & -\mathrm{A}_{21} \\ -\mathrm{A}_{12} & \mathrm{A}_{11} \end{pmatrix}}{|\mathrm{A}|}$$

Operating upon ξ with A^{-1} we find* that

$$\mathbf{R} = \begin{bmatrix} h \\ g \end{bmatrix} = -\frac{1}{2|\mathrm{A}|}\begin{pmatrix} \mathrm{A}_{22} & -\mathrm{A}_{21} \\ -\mathrm{A}_{12} & \mathrm{A}_{11} \end{pmatrix}\begin{bmatrix} d \\ e \end{bmatrix}$$

* Remember that the magnitude of a matrix A by definition is the determinant of that matrix; $|\mathrm{A}| = \text{Det}\ \mathrm{A}$.

or

$$h = -\frac{1}{2|\mathrm{A}|}(A_{22}d - A_{21}e)$$

and

$$g = -\frac{1}{2|\mathrm{A}|}(-A_{12}d + A_{11}e)$$

In a specific problem A_{11}, $A_{12} = A_{21}$, A_{22}, d and e are given; thus h and g are readily obtained. (Note that $|\mathrm{A}| = (A_{11}A_{22} - A_{12}^2)$.

Once values have been achieved for h and g the constant K can be evaluated

$$K = -\tilde{\mathbf{R}} \cdot \mathrm{A} \cdot \mathbf{R} - \tilde{\xi} \cdot \mathbf{R} - f$$

or

$$K = -[h, g]\begin{pmatrix} A_{11} & A_{12} \\ A_{21} & A_{22} \end{pmatrix}\begin{bmatrix} h \\ g \end{bmatrix} - [d, e]\begin{bmatrix} h \\ g \end{bmatrix} - f$$

$$= -A_{11}h^2 - 2A_{12}hg - A_{22}g^2 - dh - eg - f$$

It would seem that an overbearing amount of formalizing has been utilized to obtain this result. To the reader familiar with the traditional approach to quadratics in two dimensions it is apparent that these equations of translation can be achieved by elementary algebraic methods.

The matrix method actually achieves the solution with the least amount of detailed expansion in terms of components. It should be noticed that the manipulations up to the final writing of the result are purely symbolic.

This approach has the attribute that it not only works with two dimensional problems, but can be applied in exactly the same manner to problems in N dimensions. A quadratic problem in 3 or more variables can be readily solved by matrix techniques; whereas the elementary approach becomes increasingly tedious.

Properties of the Quadratic Form

The plane curves form a useful illustration of the quadratic form. In the preceding section the two dimensional quadratic was written down in its most general form.

$$A_{11}\,x_1^2 + 2A_{12}\,x_1\,x_2 + A_{22}\,x_2^2 + dx_1 + cx_2 + f = O$$

During the discussion of this problem it was mentioned that the techniques of translating to the symmetry point and those techniques to follow would apply to the **quadratic** in N dimensions,

$$\sum_{i=1}^{N}\sum_{j=1}^{N} x_i\,A_{ij}\,x_j + \sum_{j=1}^{N} \xi_j\,x_j + f = O$$

or $$\tilde{\mathbf{r}}\cdot \mathbb{A}\cdot \mathbf{r} + \tilde{\xi}\cdot \mathbf{r} + f = O$$

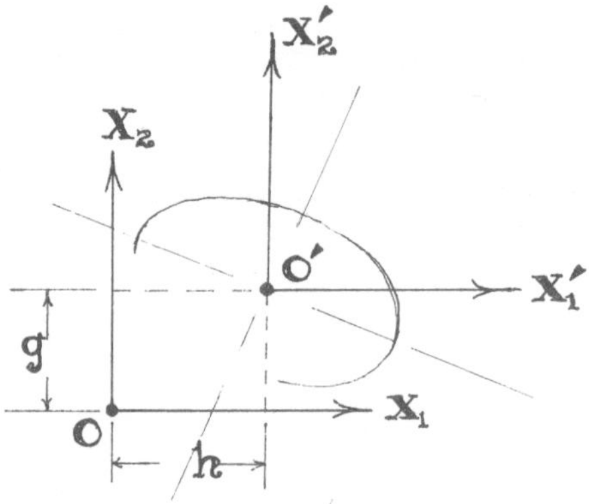

where **r** and ξ are vectors having N elements while $\mathbb{A}$ is an N x N symmetric matrix

$$A_{ij} = A_{ji}$$

In the same manner as before we can translate this equation to the symmetry point by letting

$$\mathbf{r} = \mathbf{R} + \mathbf{r}'$$

where **R** is an N dimensional constant vector.

Under these operations we find that

$$\mathbf{R} = -\frac{1}{2}\,\mathrm{A}^{-1}\cdot\xi$$

and

$$\mathrm{K} = -\tilde{\mathbf{R}}\cdot\mathrm{A}\cdot\mathbf{R} - \tilde{\xi}\cdot\mathbf{R} - \mathrm{f}$$

Substituting for **R**

$$\begin{aligned}\mathrm{K} &= \frac{1}{4}\tilde{\xi}\cdot\mathrm{A}^{-1}\cdot(\mathrm{A}\cdot\mathrm{A}^{-1})\cdot\xi + \frac{1}{2}\tilde{\xi}\cdot\mathrm{A}^{-1}\cdot\xi - \mathrm{f}\\ &= \tilde{\xi}\cdot\left\{\frac{1}{4}\widetilde{(\mathrm{A}^{-1})} + \frac{1}{2}\mathrm{A}^{-1}\right\}\cdot\xi - \mathrm{f}\end{aligned}$$

After translation we are left with an equation which involves no linear terms in **r**, (*the primes can be dropped at this point*).

$$\tilde{\mathbf{r}}\cdot\mathrm{A}\cdot\mathbf{r} = \mathrm{K}$$

or

$$\sum_{m=1}^{N}\sum_{n=1}^{N} x_m A_{mn} x_n = \mathrm{K}$$

If the reader still finds the notation bothersome expand again the two dimensional problem (i.e., set $N = 2$): it is good for the soul.

Our problem from here on will basically consist of the development of methods by which we eliminate the **off diagonal terms** or the terms involving $x_l\, x_k$, where $l \neq k$.

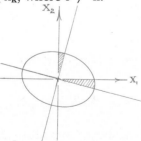

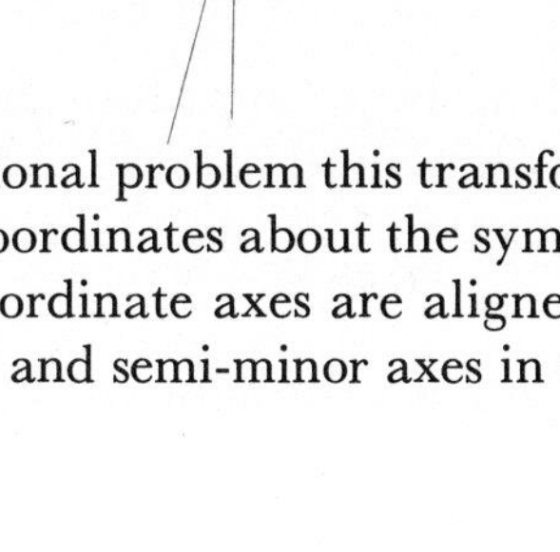

In the two dimensional problem this transformation is equivalent to a rotation of the coordinates about the symmetry point to a position in which the coordinate axes are aligned with the symmetry axes; the semi-major and semi-minor axes in the case of the ellipse.

The Discriminant

After the translation of the quadratic to the symmetry point the reader should notice that the basic matrix $\mathbb{A}$ of the quadratic is the same. Thus we say that the matrix $\mathbb{A}$ is **invariant** under the **operation** of **translation.**

Minus the magnitude of $\mathbb{A}$ in the quadratic is called the **discriminant.** Usually the discriminant is defined for the two dimensional case; however we shall extend the name to mean minus the magnitude of $\mathbb{A}$ in N dimensions.

In two dimensions, as an example, the negative of the **discriminant** is

$$|\mathbb{A}| = \begin{vmatrix} A_{11} & A_{12} \\ A_{21} & A_{22} \end{vmatrix} = A_{11}\,A_{22} - A_{12}\,A_{21} = A_{11}\,A_{22} - A_{12}{}^2$$

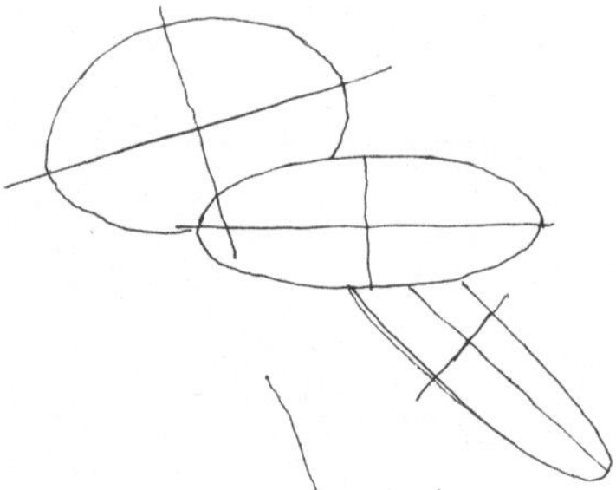

The conic sections are labeled by the properties of their magnitudes.

1. The Ellipse.

$$|\mathbb{A}| = A_{11}\,A_{22} - A_{12}{}^2 > O$$

The circle is a special case in which $A_{12} = O$ and $A_{11} = A_{22}$.

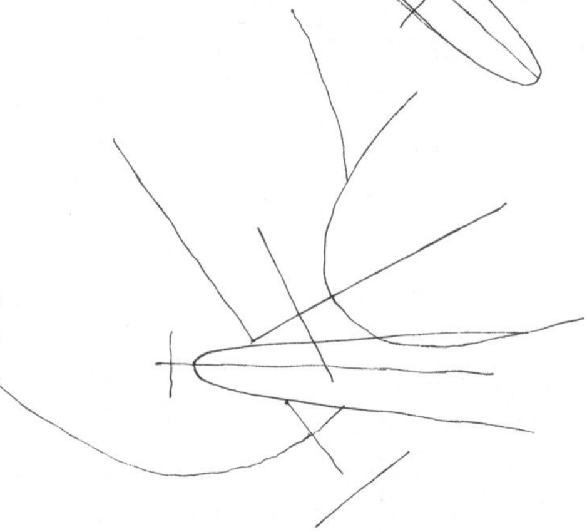

2. The Parabola.

$$|\mathbb{A}| = O$$

In this case no attempt is made to translate since either A_{11} or A_{22} is zero and A_{12} is zero.

3. The Hyperbola.

$$|\mathbb{A}| < O$$

This problem has a symmetry point. It should be noted that **two intersecting** straight lines is a special case of this form.

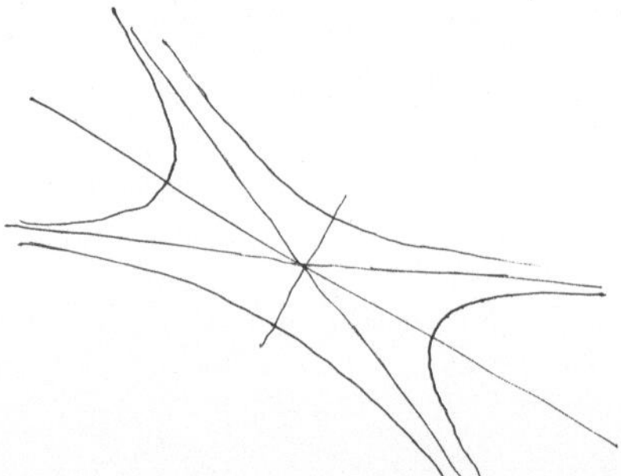

Rotation of the Quadratic Form

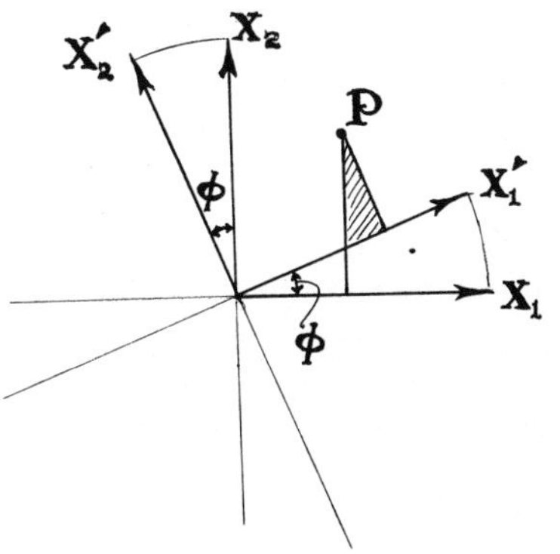

Considering the symmetric form

$$\tilde{\mathbf{r}} \cdot \mathbb{A} \cdot \mathbf{r} = [x_1, x_2] \begin{pmatrix} A_{11} & A_{12} \\ A_{21} & A_{22} \end{pmatrix} \begin{bmatrix} x_1 \\ x_2 \end{bmatrix} = K$$

or

$$A_{11}\, x_1^2 + 2A_{12}\, x_1\, x_2 + A_{22}\, x_2^2 = K$$

The rotation of the coordinates implies a **linear orthogonal transformation** of the x's,

$$x_1' = (\cos\phi)\, x_1 + (\sin\phi)\, x_2$$
$$x_2' = (-\sin\phi)\, x_1 + (\cos\phi)\, x_2$$

written

$$\mathbf{r}' = \mathbb{S} \cdot \mathbf{r} = \begin{pmatrix} \cos\phi & \sin\phi \\ -\sin\phi & \cos\phi \end{pmatrix} \begin{bmatrix} x_1 \\ x_2 \end{bmatrix}$$

In the same manner x_m can be defined in terms of the x_l' s (refer to Chapter 1)

$$x_1 = (\cos\phi)\, x_1' + (-\sin\phi)\, x_2'$$
$$x_2 = (\sin\phi)\, x_1' + (\cos\phi)\, x_2'$$

or

$$\mathbf{r} = \mathbb{S}^{-1} \cdot \mathbf{r}' = \begin{pmatrix} \cos\phi & -\sin\phi \\ \sin\phi & \cos\phi \end{pmatrix} \begin{bmatrix} x_1' \\ x_2' \end{bmatrix}$$

Although we are specifically writing out the problem of two dimensions the reader should keep in mind that the operations are in general **orthogonal transformations** and can be applied to N dimensions.

Going back to the two dimensional quadratic form

$$A_{11}\, x_1^2 + 2A_{12}\, x_1\, x_2 + A_{22}\, x_2^2 = K$$

we can substitute the expansions of x_1 and x_2 in terms of x_1' and x_2'. Upon performing this substitution we obtain a new equation

$$B_{11} x_1'^2 + 2B_{12} x_1' x_2' + B_{22} x_2'^2 = K$$

The reader should **perform this substitution** and show that in the case of the orthogonal transformation

$$|\mathbb{A}| = |\mathbb{B}|$$

or

$$A_{11} A_{22} - A_{12}^2 = B_{11} B_{22} - B_{12}^2 \qquad (1)$$

Also

$$A_{11} + A_{22} = B_{11} + B_{22} \qquad (2)$$

and

$$K = K' \qquad (3)$$

The sum of the **diagonal elements** of a **matrix** is called the **trace** and is written

$$\text{Tr}\ \mathbb{A} = A_{11} + A_{22}$$

or

$$\text{Tr}\ \mathbb{A} = \sum_{\text{all n}} A_{nn}$$

The **two important invariants** of the **quadratic** (or bilinear) form under an orthogonal transformation are

a) The magnitude of $|\mathbb{A}|$

$$\text{i.e. } |\mathbb{A}| = |\mathbb{B}|$$

b) The trace of $\mathbb{A}$

$$\text{Tr}\ \mathbb{A} = \text{Tr}\ \mathbb{B}$$

We have demonstrated these invariants in the two dimensional case and have indicated that this can be done by direct substitution of the expansions of x_l in terms of the x_k''s.

This demonstration can be performed in a much more sophisticated and **powerful** manner by utilizing our symbolic matrix notation.

Starting with

$$\tilde{\mathbf{r}} \cdot \mathbb{A}\, \mathbf{r} = \mathrm{K}$$

we substitute

$$\mathbf{r} = \mathbb{S}^{-1} \cdot \mathbf{r}', \text{ and } \tilde{\mathbf{r}} = \tilde{\mathbf{r}}' \cdot \widetilde{(\mathbb{S}^{-1})} = \tilde{\mathbf{r}}' \cdot \mathbb{S}$$

giving

$$\begin{aligned} \mathbf{r} \cdot \mathbb{A}\, \mathbf{r} = (\tilde{\mathbf{r}}' \cdot \mathbb{S}) \cdot \mathrm{A} \cdot (\mathbb{S}^{-1} \cdot \mathbf{r}') &= \tilde{\mathbf{r}}' \cdot (\mathbb{S} \cdot \mathrm{A} \cdot \mathbb{S}^{-1}) \cdot \mathbf{r}' \\ &= \tilde{\mathbf{r}}' \cdot \mathbb{B} \cdot \mathbf{r}' = \mathrm{K} \end{aligned}$$

Then

$$\mathbb{S} \cdot \mathbb{A} \cdot \mathbb{S}^{-1} = \mathbb{B}$$

This orthogonal transformation of $\mathbb{A}$ is called a **similarity transformation.**

The **invariants** under a similarity transformation can be demonstrated quite readily.

1. Invariance of $|\mathbb{A}|$

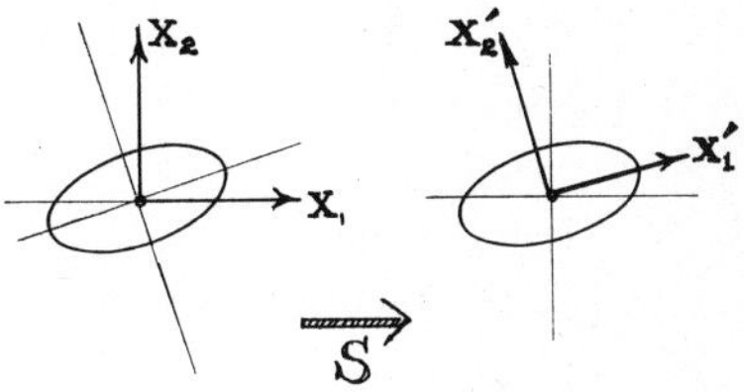

Note that

$|\mathbb{B}| = |\mathbb{S} \cdot \mathbb{A} \cdot \mathbb{S}^{-1}| = |\mathbb{S}||\mathbb{A}||\mathbb{S}^{-1}|$

Because the magnitude of an orthogonal matrix is one;

$$|\mathbb{S}| = 1 \text{ and } |\mathbb{S}^{-1}| = 1$$

we have the result

$$|\mathbb{B}| = |\mathbb{A}|$$

2. Invariance of $\mathrm{Tr}\,\mathbb{A} = \sum_{n=1}^{N} \mathrm{A}_{nn}$

The elements of $\mathbb{S} \cdot \mathbb{A} \cdot \mathbb{S}^{-1}$ are

$$\mathrm{B}_{mn} = (\mathbb{S} \cdot \mathbb{A} \cdot \mathbb{S}^{-1})_{mn} = \sum_{k=1}^{N} \sum_{l=1}^{N} \mathrm{S}_{mk}\, \mathrm{A}_{kl}\, \mathrm{S}_{ln}'$$

Consider

$$\mathrm{Tr}\,\mathbb{B} = \sum_{m=1}^{N} \mathrm{B}_{mm} = \sum_{m=1}^{N} \sum_{k=1}^{N} \sum_{l=1}^{N} \mathrm{S}_{mk}\, \mathrm{A}_{kl}\, \mathrm{S}_{lm}'$$

If we sum over m first remembering that $\mathbb{S} \cdot \mathbb{S}^{-1} = \mathbb{I}$ or $\sum_m S_{mk} S_{lm}' = \delta_{kl}$,

we find

$$\text{Tr } \mathbb{B} = \sum_{k=1}^{N}\sum_{l=1}^{N} A_{kl} \{\sum_{m=1}^{N} S_{lm}' S_{mk}\} = \sum_{k=1}^{N}\sum_{l=1}^{N} A_{kl}\, \delta_{lk}$$

$$\text{Tr } \mathbb{B} = \sum_{k=1}^{N} A_{kk} = \text{Tr } \mathbb{A}$$

Again! if the reader finds this too elegant, perform the **sum directly** for N = 2, i.e., for a two dimensional case.

Diagonalization of the Quadratic

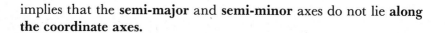

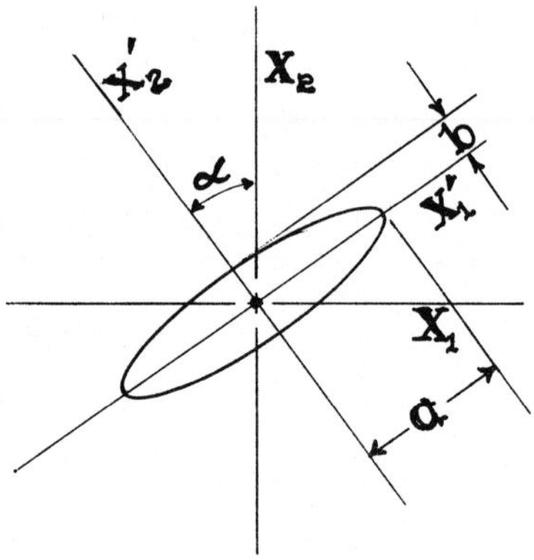

In two dimensions the presence of the term

$$2A_{12}\, x_1\, x_2$$

in the quadratic

$$\tilde{\mathbf{r}} \cdot \mathbb{A} \cdot \mathbf{r} = K$$

implies that the **semi-major** and **semi-minor** axes do not lie **along the coordinate axes.**

Starting with a general quadratic one of the most useful problems is that of finding that particular orthogonal transformation (rotation) which makes the quadratic **diagonal,** in other words we rotate to eliminate all terms of the form $2A_{lm}\, x_l\, x_m$ where $l \neq m$.

Associated with this we also want to discover the lengths of the semi-major and semi-minor axes. These lengths are called the **eigenvalues** of the **quadratic.**

We will first obtain the rotation to normal form by the technique most often described in elementary texts. In the following section we will employ a more powerful and more convenient method using **matrices.**

The simplest method of relating the normal form to the general form is by writing the normal form in x_1', x_2' and rotating through an angle α to the general form.

The normal form is

$$\frac{x_1'^2}{a^2} + \frac{x_2'^2}{b^2} = 1$$

We can relate x_1' and x_2' to x_1 and x_2 by means of the equations of rotation (or the rotation matrix).

$$\mathbf{r}' = \begin{bmatrix} x_1' \\ x_2' \end{bmatrix} = \mathbb{S} \cdot \mathbf{r}$$

giving

$$x_1' = (\cos\alpha)\, x_1 + (\sin\alpha)\, x_2$$
$$x_2' = (-\sin\alpha)\, x_1 + (\cos\alpha)\, x_2$$

Substituting for x_1' and x_2' in the **normal form,** we obtain

$$Ax_1^2 + Bx_1x_2 + Cx_2^2 = 1^*$$

where

$$A = \frac{\cos^2\alpha}{a^2} + \frac{\sin^2\alpha}{b^2}$$

$$B = 2\frac{1}{a^2} - \frac{1}{b^2}\cos\alpha\sin\alpha$$

$$C = \frac{\sin^2\alpha}{a^2} + \frac{\cos^2\alpha}{b^2}$$

* Here we have used $A_{11} = A$, $2A_{12} = B$ and $A_{22} = C$.

The **discriminant is invariant,**

$$B^2 - 4AC = -4\,\frac{\cos^2\alpha + \sin^2\alpha}{a^2b^2} = -\frac{4}{a^2b^2}$$

(does not depend on α).

In addition to this

$$A + C = \frac{1}{a^2} + \frac{1}{b^2}$$

The angle of rotation

$$\cos(2\alpha) = \frac{C - A}{B}$$

Diagonalization of the General Form

The diagonalization of the **quadratic in N dimensions** can be obtained by more general techniques. Once again the two dimensional quadratic will be utilized to illustrate the general method.

Starting with the **non-diagonal form**

$$\tilde{\mathbf{r}} \cdot \mathbb{A} \cdot \mathbf{r} = [x_1, x_2]\begin{pmatrix} A_{11} & A_{12} \\ A_{21} & A_{22} \end{pmatrix}\begin{bmatrix} x_1 \\ x_2 \end{bmatrix} =$$

$$A_{11}x_1^2 + 2A_{12}x_1x_2 + A_{22}x_2^2 = K$$

We want the transformation (rotation in two dimensions) $\mathbb{S}$ which produces the **diagonal form,** Γ* where;

$$\tilde{\mathbf{r}}' \cdot \Gamma \cdot \mathbf{r}' = [x_1', x_2']\begin{pmatrix} \Gamma_{11} & O \\ O & \Gamma_{22} \end{pmatrix}\begin{bmatrix} x_1' \\ x_2' \end{bmatrix} =$$

$$\Gamma_{11}\, x_1'^2 + \Gamma_{22}\, x_2'^2 = K$$

The orthogonal transformation from the **form** $\mathbb{A}$ to the **diagonal form** Γ is obtained by operator $\mathbb{S}$; where

$$\mathbf{r} = \mathbb{S}^{-1} \cdot \mathbf{r}', \text{ or } \tilde{\mathbf{r}} = \mathbf{r}' \cdot \mathbb{S}$$

* Γ is a special case of the transformed matrix B discussed in section F-2.

Substituting for **r** we obtain

$$\tilde{\mathbf{r}} \cdot \mathbb{A} \cdot \mathbf{r} = \tilde{\mathbf{r}}' \cdot (\mathbb{S} \cdot \mathbb{A} \cdot \mathbb{S}^{-1}) \cdot \mathbf{r}' = \tilde{\mathbf{r}}' \cdot \Gamma \cdot \mathbf{r}' = K$$

As before under the condition that $\mathbb{S}$ is an **orthogonal transformation**

$$\mathbb{S} \cdot \mathbb{A} \cdot \mathbb{S}^{-1} = \Gamma$$

or

$$\begin{pmatrix} S_{11} & S_{12} \\ S_{21} & S_{22} \end{pmatrix} \begin{pmatrix} A_{11} & A_{12} \\ A_{21} & A_{22} \end{pmatrix} \begin{pmatrix} S_{11}' & S_{12}' \\ S_{21}' & S_{22}' \end{pmatrix} = \begin{pmatrix} \Gamma_{11} & O \\ O & \Gamma_{22} \end{pmatrix}$$

Several things must be kept in mind;

$$\mathbb{S}^{-1} = \tilde{\mathbb{S}}$$

$$\mathbb{S}^{-1} = \begin{pmatrix} S_{11} & S_{21} \\ S_{12} & S_{22} \end{pmatrix}$$

second, while $\mathbb{S}$ is not symmetric, always remember that $\mathbb{A}$ is symmetric

$$A_{12} = A_{21}$$

To accomplish the diagonalization of $\mathbb{A}$ we utilize the basic characteristic of $\mathbb{S}$, namely that the **magnitude** of **r** is **invariant under** the **transformation** $\mathbb{S}$.

$$\tilde{\mathbf{r}} \cdot \mathbf{r} = \tilde{\mathbf{r}}' \cdot \mathbf{r}'$$

We multiply $|\mathbf{r}|^2$ by some constant λ so that the product equals the constant K; then subtract this equation from the quadratic,

$$\tilde{\mathbf{r}} \cdot \mathbb{A} \cdot \mathbf{r} = \tilde{\mathbf{r}}' \cdot \Gamma \cdot \mathbf{r}' = K$$

Subtracting

$$\lambda\, \tilde{\mathbf{r}} \cdot \mathbb{I} \cdot \mathbf{r} =$$

$$= \lambda\, \tilde{\mathbf{r}}' \cdot \mathbb{I} \cdot \mathbf{r}' = K$$

we obtain

$$\tilde{\mathbf{r}} \cdot \{\mathbb{A} - \lambda \mathbb{I}\} \cdot \mathbf{r} = \tilde{\mathbf{r}}' \cdot \{\Gamma - \lambda \mathbb{I}\} \cdot \mathbf{r}' = O$$

At this stage in the development of the analysis of quadratics a similarity transformation of the symmetric matrix $\mathbb{A}$ has been made taking it to another (congruent) matrix Γ. Although it was claimed that there exists one rotation which makes Γ diagonal, the proof was not given.

Using the 2 x 2 matrix as an example for simplicity we can demonstrate that $\mathbb{A}$ can be diagonalized and simultaneously that the actual algebraic operations can be represented graphically. To accomplish these results let us return to the general quadratic,

$$\tilde{\mathbf{r}} \cdot \mathbb{A} \cdot \mathbf{r} = \tilde{\mathbf{r}}' \cdot \Gamma \cdot \mathbf{r}' = 1$$

From here on K will be taken as 1 without any loss of generality.

Both of these bilinear forms can be thought of as the equations for an ellipse.

The final quadratic representing the equation for **two** straight lines was obtained by subtracting the bilinear circular forms from the elliptic forms. The circular forms were

$$\lambda \, \tilde{\mathbf{r}} \cdot \mathbb{I} \cdot \mathbf{r} = \lambda \, \tilde{\mathbf{r}}' \cdot \mathbb{I} \cdot \mathbf{r}' = 1$$

where λ equals one over the square of the radius of the circle;

$$\lambda = \frac{1}{R^2}$$

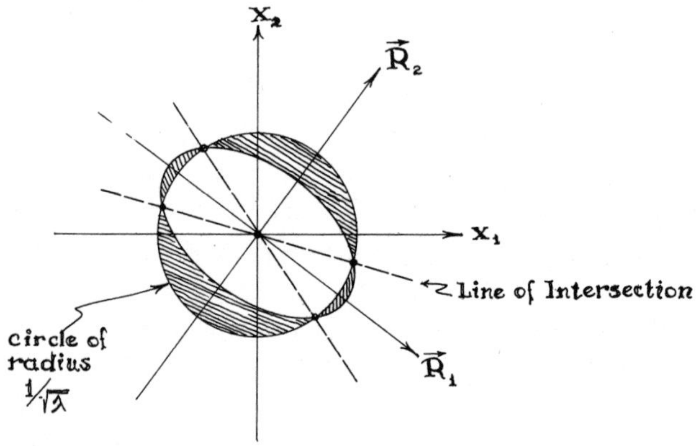

The subtraction represents a solution which gives **two straight lines** passing through the points of intersection of the circle and the

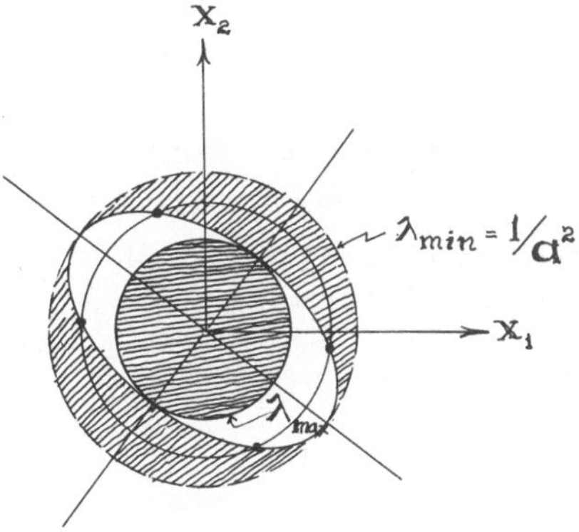

ellipse. The geometric construction of $\tilde{\mathbf{r}} \cdot (\mathbb{A} - \lambda \mathbb{I}) \cdot \mathbf{r} = 0$ indicates that there is a maximum and minimum value of $R = \frac{1}{\sqrt{\lambda}}$ resulting from the two conditions under which the circle is just tangent to the ellipse.

The existence of the upper and lower limit upon λ can be demonstrated by solving the quadratic,

$$\tilde{\mathbf{r}} \cdot (\mathbb{A} - \lambda \mathbb{I}) \cdot \mathbf{r} = (A_{11} - \lambda)\, x_1^2 + 2A_{12}\, x_1\, x_2 + (A_{22} - \lambda)\, x_2^2 = 0$$

for x_1 in terms of x_2 and by utilizing the fact that both x_1 and x_2 must be **real.** Solving for x_1

$$x_1 = \frac{\{-A_{12} \pm \sqrt{A_{12}^2 - (A_{11} - \lambda)(A_{22} - \lambda)}\} x_2.}{(A_{11} - \lambda)}$$

or

$$x_1 = \frac{\{-A_{12} \pm \sqrt{-\,|\mathbb{A} - \lambda \mathbb{I}|}\}\, x_2}{(A_{11} - \lambda)}$$

The condition that x_1 and x_2 be **real** requires that

$$A_{12}^2 - (A_{11} - \lambda)(A_{22} - \lambda) \geq 0$$

Therefore the magnitude of $(\mathbb{A} - \lambda\mathbb{I})$ must be negative definite;

$$|\mathbb{A} - \lambda\mathbb{I}| \leq O$$

From the relation between x_1 and x_2 we find that*

$$\lambda_{min} \leq \lambda \leq \lambda_{max}$$

The proof of this statement is apparent from the geometric example. Algebraic proof is left as an exercise.

The minimum value of λ occurs when the radius of the circle is equal to the length of the semi-major axis

$$R = a = 1/\sqrt{\lambda_{min}}$$

In like manner the maximum value, λ_{max}, is obtained when the radius of the circle is equal to the length of the semi-minor axis

$$R = b = 1/\sqrt{\lambda_{max}}$$

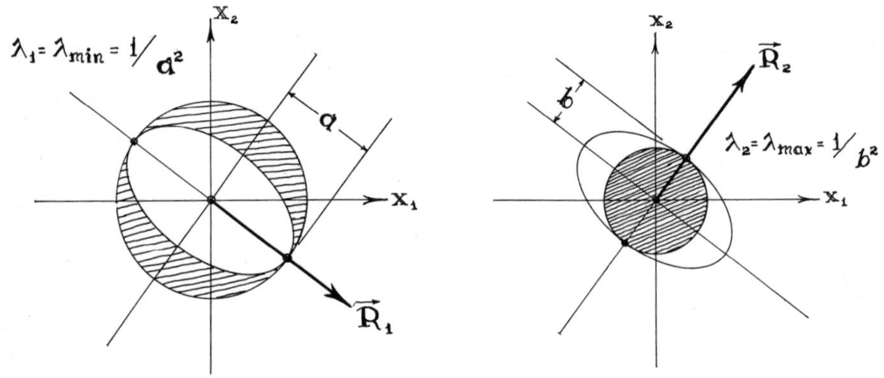

* We have utilized the ellipse to illustrate this algebraic procedure. Later the problem of the hyperbola will also be analyzed as an eigenvalue problem. In this case the approach is exactly the same algebraically. The graphical analysis utilizes tangency to the primary hyperbola and to its conjugate in order to set the maximum and minimum λ. The reader should demonstrate as an exercise that the circle tangent to the conjugate hyperbola is associated with a *negative eigenvalue.*

From the pictures of the maximum and minimum condition on λ we see that the two straight lines defined by the intersection of the circle and the ellipse coalese into one line when λ takes on either its maximum value or its minimum value. These lines are observed to lie along the principle axes of the quadratic, i.e., the semi-major or semi-minor axes.

Because

$$x_1 = \frac{\{-A_{12} \pm \sqrt{-\,|\mathbb{A} - \lambda \mathbb{I}|}\}}{(A_{11} - \lambda)} x_2$$

or

$$x_2 = \frac{\{-A_{12} \pm \sqrt{-\,|\mathbb{A} - \lambda \mathbb{I}|}\}}{(A_{22} - \lambda)} x_1$$

a single line solution is obtained only when $|\mathbb{A} - \lambda \mathbb{I}|$ takes on its **maximum value**

$$|\mathbb{A} - \lambda \mathbb{I}| = O$$

The determinant of $(\mathbb{A} - \lambda \mathbb{I})$ is known as the **secular determinant** while

$$|\mathbb{A} - \lambda \mathbb{I}| = (A_{11} - \lambda)(A_{22} - \lambda) - A_{12}^2 = O$$

is called the **characteristic equation.**

Here the characteristic equation is a polynomial of second degree and it has two roots λ_1 and λ_2 called the **eigenvalues.**

$$\lambda_j = \frac{(A_{11} + A_{22})}{2} \pm \sqrt{\frac{(A_{11} - A_{22})^2}{4} + A_{12}^2}$$

In general for a space of N dimensions the characteristic equation is a polynomial of Nth degree having N roots or **eigenvalues.**

Let λ_j be any one of the eigenvalues or roots. Because

$$|\mathbb{A} - \lambda \mathbb{I}| = |\Gamma - \lambda \mathbb{I}| = (\lambda_1 - \lambda)(\lambda_2 - \lambda) = O$$

it is possible to find one rotation S which makes

$$\Gamma = \begin{pmatrix} \lambda_1 & O \\ O & \lambda_2 \end{pmatrix} = \begin{pmatrix} \Gamma_{11} & O \\ O & \Gamma_{11} \end{pmatrix}$$

λ_j can now be utilized in the remaining discussion in place of Γ_{jj}.

If this is the case

$$\begin{vmatrix} (\lambda_1 - \lambda) & O \\ O & (\lambda_2 - \lambda) \end{vmatrix} = |\Gamma - \lambda I| = (\lambda_1 - \lambda)(\lambda_2 - \lambda) = O$$

which satisfies the condition on the roots of the polynomial. For the case of N dimensions and N roots λ_j,

$$|A - \lambda I| = |\Gamma - \lambda I| = \prod_{j=1}^{N} (\lambda_j - \lambda) = O$$

where $\prod_{j=1}^{N}$ signifies a repeated product,

$$\prod_{j=1}^{N} (\lambda_j - \lambda) = (\lambda_1 - \lambda)(\lambda_2 - \lambda) \ldots (\lambda_N - \lambda).$$

The previous discussion shows that a symmetric matrix A can be diagonalized by a similarity transformation* and that the elements of the congruent diagonal matrix Γ are obtained as roots of the characteristic equation.

Further these roots λ_j are equal to one over the square of the lengths of the principle axes.

* We can use this transformation to demonstrate that A can be diagonalized. $S \cdot A \cdot S^{-1} = \Gamma$ can be written $A \cdot S^{-1} = S^{-1} \cdot \Gamma$, or $\sum_n A_{mn} S_{nj}' = \sum_n S_{mn}' \Gamma_{nj}$. Assume that Γ is diagonal then $\sum_n A_{mn} S_{nj}' = S_{mj}' \Gamma_{jj}$, or $\sum_n \{A_{mn} - \Gamma_{jj} \delta_{mn}\} S_{nj}' = 0$. A sufficient condition that this last equation have a nontrivial solution is that $|A - \Gamma_{jj} I| = O$. The existence of a solution demonstrates that Γ can be diagonalized, keeping in mind that S^{-1} is orthogonal. This can be accomplished if A is symmetric.

This is made more obvious by examining the ellipse $\tilde{\mathbf{r}}' \cdot \Gamma \cdot \mathbf{r}' = 1$.

$$\tilde{\mathbf{r}}' \cdot \Gamma \cdot \mathbf{r}' = [x_1', x_2'] \begin{pmatrix} \lambda_1 & O \\ O & \lambda_2 \end{pmatrix} \begin{bmatrix} x_1' \\ x_2' \end{bmatrix}$$

$$= \lambda_1 x_1'^2 + \lambda_2 x_2'^2 = \frac{x_1'^2}{a^2} + \frac{x_2'^2}{b^2} = 1$$

In other words $\tilde{\mathbf{r}}' \cdot \Gamma \cdot \mathbf{r}' = 1$ is the **normal form.**

The directions of the principle axes (semi-major and semi-minor in two dimensions) can be developed by utilizing the condition $|A - \lambda I| = O$.

If we impose the condition on the vector **r** that the linear form

$$\{A - \lambda I\} \cdot \mathbf{r} = O$$

then a necessary condition that the solutions for the components of **r** be **non-trivial** (i.e., not all zero) is that

$$|A - \lambda I| = O$$

This condition of course is that constraint which gives the **eigenvalues** λ_j as the roots of this equation. If these **unique values** or **eigenvalues** are then employed in the linear form, the solutions for the components of the corresponding vector are vectors *parallel to the principle axes.*

$$\{A - \lambda_j I\} \cdot \mathbf{R}_j = O$$

The constraint that this linear form vanish is a sufficient (not necessary) condition that the quadratic relationship be satisfied. In other words

$$\tilde{\mathbf{R}}_j \cdot \{A - \lambda_j I\} \cdot \mathbf{R}_j; = O$$

since $\mathbf{R}_j$ times a "null" vector is zero.

The solutions $\mathbf{R}_j$ (having components X_{mj}) of

$$\{A - \lambda_j I\} \cdot \mathbf{R}_j = \sum_{m=1}^{N} A_{lm} X_{mj} - \delta_{lm} X_{mj} \lambda_j = O$$

are called the **eigenvectors** of the problem. We changed the notation from **r** to $\mathbf{R}_j$ because of the restricted nature of the **R**'s. They satisfy the linear eigenvalue equation in a non-trivial fashion.

In the earlier portion of this section the *characteristic equation* was required in order that we obtain the single straight line solutions to the quadratic. We remarked at that time that these lines passed through the points of tangency between the circle and the ellipse.

The **eigenvectors** are vector forms of those same lines because of their unique dependence upon the **eigenvalues.** More will be said concerning the eigenvectors in the next section. However, before proceeding let us restate the important consequences of this discussion.

1. The **diagonalization** of the quadratic is performed by finding the roots λ_j of the polynomial $|A - \lambda I| = O$.
These roots are called the eigenvalues. The length of the jth principle axis is one over the square root of λ_j.

2. The direction of the jth principle axis is determined by using λ_j in the eigenvalue problem $(A - \lambda_j I) \cdot \mathbf{R}_j = O$.
The solutions $\mathbf{R}_j$ are called the eigenvectors and are directed along the principle axes.

Before closing this section one unique quadratic, the circle, should be mentioned. This quadratic is degenerate in that the two eigenvalues are the same. Whenever a quadratic has degenerate eigenvalues, the associated eigenvectors are not specifically determined in spite of the fact that they remain mutually perpendicular. To understand this the reader should consider the two dimensional problem of the circle. Any axis serves as say a semi-major axis. However, once one axis is chosen (arbitrarily) the second axis is fixed because it must be perpendicular to the first.

To demonstrate the calculation of eigenvalues consider an ellipse which has its semi-major axis at an angle of $-30°$ relative to the x_1 axis

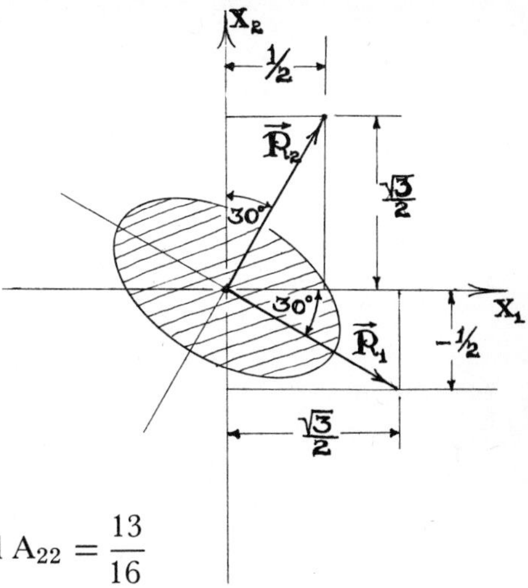

$$\frac{7}{16}x_1^2 + \frac{3\sqrt{3}}{8}x_1x_2 + \frac{13}{16}x_2^2 = 1$$

In this problem

$$A_{11} = 7/16,\ A_{12} = A_{21} = \frac{3\sqrt{3}}{16},\ \text{and } A_{22} = \frac{13}{16}$$

Thus

$$\mathbb{A} = \begin{pmatrix} 7/16 & \dfrac{3\sqrt{3}}{16} \\ \dfrac{3\sqrt{3}}{16} & \dfrac{13}{16} \end{pmatrix}$$

and

$$|\mathbb{A} - \lambda\mathbb{I}| = \begin{vmatrix} (7/16 - \lambda) & \dfrac{3\sqrt{3}}{16} \\ \dfrac{3\sqrt{3}}{16} & \left(\dfrac{13}{16} - \lambda\right) \end{vmatrix} = O$$

When we expand the determinant and solve for λ we find

$$(7/16 - \lambda)\left(\frac{13}{16} - \lambda\right) - \frac{27}{(16)^2} = O$$

with the solutions

$$\lambda_1 = 5/8 - 3/8 = \frac{1}{4} = \frac{1}{a^2}$$

and

$$\lambda_2 = 5/8 + \frac{3}{8} = 1 = \frac{1}{b^2}$$

The equation

$$\frac{x_1'^2}{4} + \frac{x_2'^2}{1} = 1 \text{ is in the normal form.}$$

Another simple example of the problem of diagonalization is that of the hyperbola

$$x_1 x_2 = 1$$

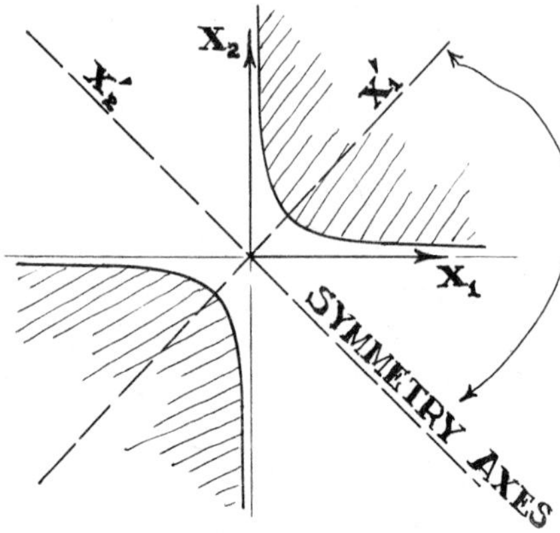

The eigenvalues give one over the squares of the distances from the origin to the hyperbola and to its conjugate. As in the case of the ellipse the problem has a set of symmetry axes. In solving this problem we will find that one of the fundamental lengths would be imaginary because one of the eigenvalues of the problem is negative. This is the eigenvalue associated with the conjugate hyperbola.

In the problem above

$$A_{11} = O, A_{12} = A_{21} = 1/2 \text{ and } A_{22} = O$$

Then

$$|A - \lambda I| = \begin{vmatrix} -\lambda & 1/2 \\ 1/2 & -\lambda \end{vmatrix} = O$$

The two roots are

$$\lambda_1 = \Gamma_{11} = \frac{1}{a^2} = + 1/2$$

and

$$\lambda_2 = \Gamma_{22} = \frac{1}{b^2} = - 1/2$$

Thus $\frac{x_1'^2}{2} - \frac{x_2'^2}{2} = 1$ is the normal form.

The Eigenvalue Problem

When the constant K of a bilinear or quadratic form is equal to zero the equation is degenerate in that it defines

two straight lines

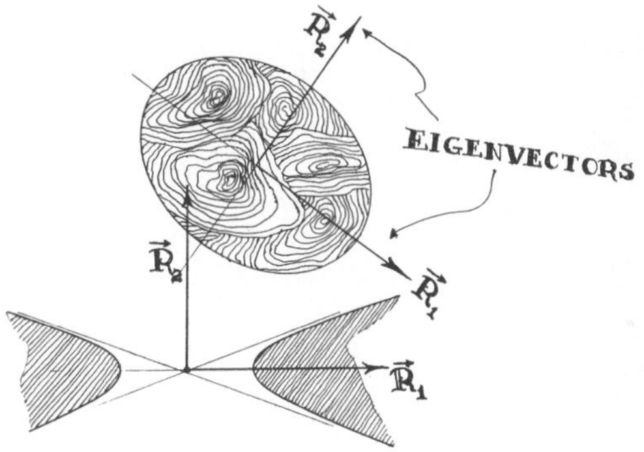

Under such conditions the quadratic can be factored into the product of a set of **linear algebraic equations.**

Stating this in mathematical language, the equation,

$$\tilde{\mathbf{r}} \cdot \mathbb{C} \cdot \mathbf{r} = [x_1, x_2] \begin{pmatrix} C_{11} & C_{12} \\ C_{21} & C_{22} \end{pmatrix} \begin{bmatrix} x_1 \\ x_2 \end{bmatrix} = O$$

is the equation of two straight lines. To show this expand the equation above

$$\mathbf{r} \cdot \mathbb{C} \cdot \mathbf{r} = C_{11}\, x_1^2 + 2C_{12}\, x_1x_2 + C_{22}\, x_2^2 = O$$

This equation can be factored into the form

$$(x_1 - mx_2)\,(x_1 - nx_2) = O$$

The relation above has two solutions;

$$x_1 = mx_2$$

or

$$x_1 = nx_2$$

In the previous section it was shown that in the case of a circle subtracted from an ellipse, these two lines passed through the points of intersection of the circle and the ellipse.

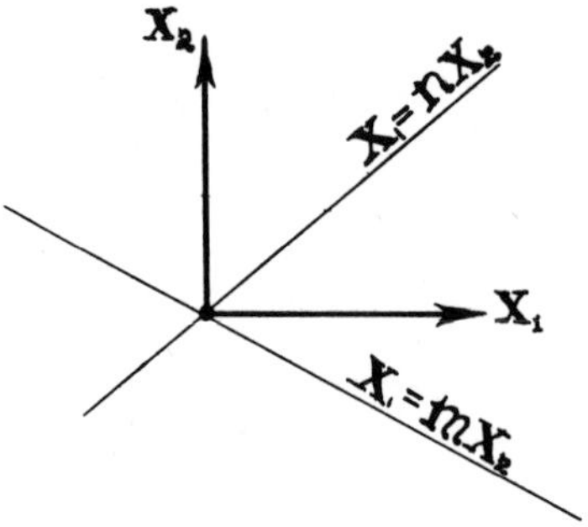

The diagonalization of the quadratic was achieved by creating a new quadratic which had the constant term **equal to zero.**

$$\tilde{\mathbf{r}} \cdot \{\mathbb{A} - \lambda \mathbb{I}\} \cdot \mathbf{r} = 0$$

The extreme cases of this equation occur when only one line is defined, in other words the lines through the points of intersection coalese into one line and the circle is tangent to the quadratic $\mathbb{A}$. When only one line is defined the magnitude of $(\mathbb{A} - \lambda \mathbb{I})$ takes on its maximum value, zero. Under this condition particular values of λ are obtained, and these are called the eigenvalues. As we have shown the **eigenvalues** are the **roots** of the **characteristic equation**

$$|\mathbb{A} - \lambda \mathbb{I}| = 0$$

We labeled these roots $\lambda_1, \lambda_2, \ldots \lambda_N$, or $\Gamma_{11}, \Gamma_{22}, \ldots \Gamma_{NN}$.

Consider again the equation

$$\mathbf{r} \cdot \{\mathbb{A} - \lambda \mathbb{I}\} \cdot \mathbf{r} = O$$

If this equation is zero it is sufficient that

$$\{\mathbb{A} - \lambda_j \mathbb{I}\} \cdot \mathbf{R}_j = O$$

where $\mathbf{R}_j$ is a vector lying along the jth symmetry axis.
In order to investigate the details of the calculation of the $\mathbf{R}_j$'s call the two roots $\lambda_1 = \frac{1}{a^2}$ and $\lambda_2 = \frac{1}{b^2}$. Each specific root then determines a specific vector $\mathbf{R}_j$ called the **eigenvector** corresponding to the **eigenvalue** λ_j.

Write

$$\mathbf{R}_j = \begin{bmatrix} X_{1j} \\ X_{2j} \end{bmatrix}$$

Then,

$$\{\mathbb{A} - \lambda_j \mathbb{I}\} \cdot \mathbf{R}_j = \begin{pmatrix} (A_{11} - \lambda_j) & A_{12} \\ A_{21} & (A_{22} - \lambda_j) \end{pmatrix} \begin{bmatrix} X_{1j} \\ X_{2j} \end{bmatrix} = O$$

This matrix multiplication gives us *two* algebraic equations specifying the ratio of the components of $\mathbf{R}_j$.

$$\mathbb{A} \cdot \mathbf{R}_j = \lambda_j \mathbf{R}_j$$

or

$$A_{11} X_{1j} + A_{12} X_{2j} = \lambda_j X_{1j}$$
$$A_{21} X_{1j} + A_{22} X_{2j} = \lambda_j X_{2j}$$

These two equations are not independent in that they give the same ratio of X_{1j} to X_{2j}.

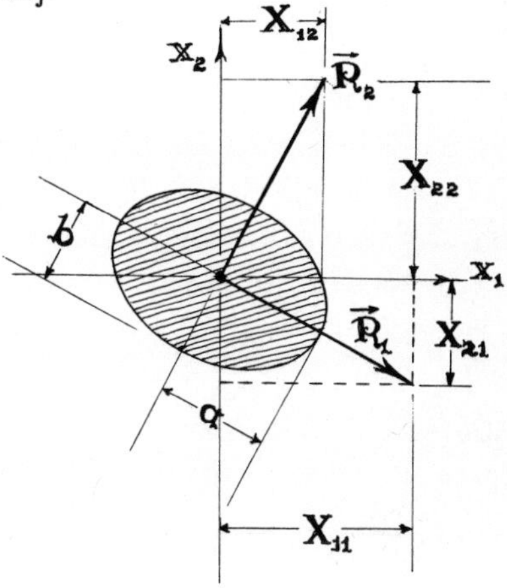

$$\frac{X_{1j}}{X_{2j}} = -\frac{A_{12}}{A_{11} - \lambda_j}$$

or (Both forms provide the same result.)

$$\frac{X_{1j}}{X_{2j}} = -\frac{A_{22} - \lambda_j}{A_{21}}$$

The length of the vector can be set by normalizing $\mathbf{R}_j$ (setting $|\mathbf{R}_j| = 1$). $\tilde{\mathbf{R}}_j \cdot \mathbf{R}_j = X_{1j}^2 + X_{2j}^2 = 1$.

$$X_{1j}^2 \left\{1 + \frac{(A_{11} - \lambda_j)^2}{(A_{12})^2}\right\} = 1$$

and

$$X_{1j}^2 = \frac{(A_{12})^2}{(A_{11} - \lambda_j)^2 + (A_{12})^2}$$

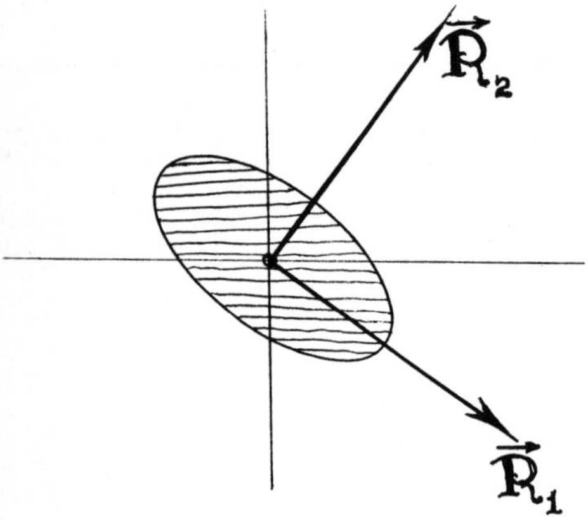

In general the eigenvectors (or principle axes) of a **symmetric matrix** are **orthogonal.** (Symmetric, $A_{jk} = A_{kj}$). It is apparent that many matrices have eigenvalues and eigenvectors; however, only the symmetric matrices will always have orthogonal eigenvectors.

In the preceding section concerning diagonalization of the quadratic we solved for the eigenvalues of the ellipse

$$\frac{7}{16}x_1^2 + \frac{3\sqrt{3}}{8}x_1x_2 + \frac{13}{16}x_2^2 = 1$$

and obtained

$$\lambda_1 = \frac{1}{a^2} = \frac{1}{4}$$

and

$$\lambda_2 = \frac{1}{b^2} = 1$$

To obtain the direction of the symmetry axes, we must solve the eigenvalue problem

$$\{\mathbb{A} - \lambda_j \mathbb{I}\} \cdot \mathbf{R}_j = O$$

or

$$\begin{pmatrix} \left(\frac{7}{16} - \lambda_j\right) & \frac{3\sqrt{3}}{16} \\ \frac{3\sqrt{3}}{16} & \left(\frac{13}{16} - \lambda_j\right) \end{pmatrix} \begin{bmatrix} X_{1j} \\ X_{2j} \end{bmatrix} = O$$

Take

$$\lambda_1 = 1/4\ (j = 1) \text{ then}$$

$$\left(\frac{7}{16} - \frac{1}{4}\right) X_{11} + \frac{3\sqrt{3}}{16} X_{21} = O \qquad X_{21} = \frac{-1}{\sqrt{3}} X_{11}$$

Normalizing we obtain,

$$X_{11}^{\ 2} + X_{21}^{\ 2} = \frac{4}{3} X_{11}^{\ 2} = 1, \text{ and } X_{11} = \frac{\sqrt{3}}{2}$$

Giving in addition

$$X_{21} = -\frac{1}{2}$$

Now take $j = 2$ or $\lambda_2 = 1$, then

$$\left(\frac{7}{16} - 1\right) X_{12} + \frac{3\sqrt{3}}{16} X_{22} = O \qquad X_{12} = \frac{1}{\sqrt{3}} X_{22}$$

Normalizing

$$X_{12}^2 + X_{22}^2 = X_{22}^2 \left\{1 + \frac{1}{3}\right\} = 1; X_{22} = \frac{\sqrt{3}}{2}$$

and $$X_{12} = \frac{1}{2}$$

We can see this result graphically.

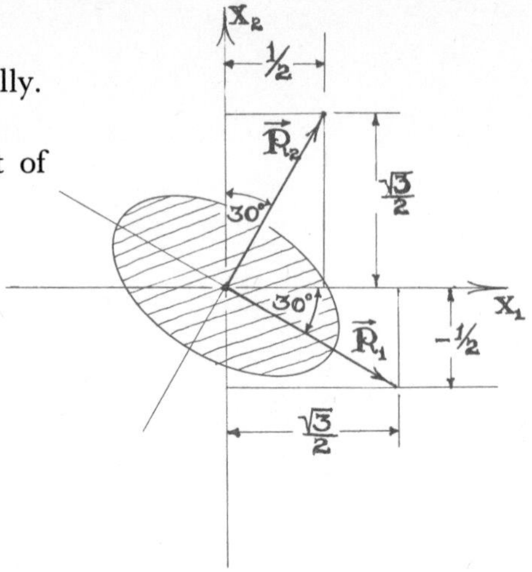

Our other example was that of the hyperbola

$$x_1 x_2 = 1$$

with

$$\lambda_1 = \frac{1}{2}$$

and

$$\lambda_2 = -\frac{1}{2}$$

Using these eigenvalues we can readily find the **eigenvectors.**

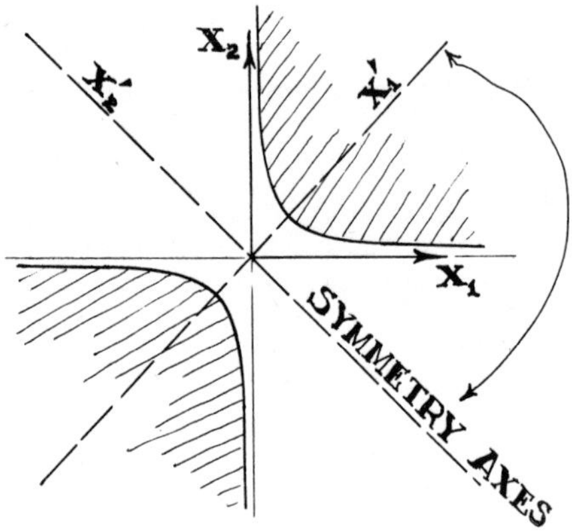

$$\{\mathbb{A} - \lambda_j \mathbb{I}\} \cdot \mathbf{R}_j = O$$

or

$$\begin{pmatrix} -\lambda_j & \frac{1}{2} \\ \frac{1}{2} & -\lambda_j \end{pmatrix} \begin{bmatrix} X_{1j} \\ X_{2j} \end{bmatrix} = O$$

Take

$$j = 1; \lambda_1 = \frac{1}{2}$$

Then

$$-\frac{1}{2} X_{11} + \frac{1}{2} X_{21} = O$$

or

$$X_{11} = X_{21}$$

After normalizing

$$\mathbf{R}_1 = \frac{1}{\sqrt{2}} \begin{bmatrix} 1 \\ 1 \end{bmatrix}, \text{ or } \mathbf{R}_1 = \frac{1}{\sqrt{2}} \{\epsilon_1 + \epsilon_2\}$$

Taking $$j = 2; \lambda_2 = -\frac{1}{2}$$

and $$X_{12} = -X_{22} = -\frac{1}{2}$$

Thus $$\mathbf{R}_2 = \frac{1}{\sqrt{2}}\begin{bmatrix} -1 \\ \\ 1 \end{bmatrix}, \text{ or } \mathbf{R}_2 = \frac{1}{\sqrt{2}}\{-\epsilon_1 + \epsilon_2\}$$

The sign is arbitrary for this vector. The sign was taken in order that $\mathbf{R}_1$, $\mathbf{R}_2$ form a **right handed system,** (or that $(\mathbf{R}_1 \times \mathbf{R}_2)$ is positive).

Properties of the Eigenvectors

The relation between $\mathbf{R}_j$ and $\mathbb{S}^{-1}$.

It can be demonstrated in a relatively simple manner that the elements of the *normalized* $\mathbf{R}_j$ form the ROWS in the $\mathbb{S}^{-1}$ rotation matrix.

The eigenvalue problem is

$$\{\mathbb{A} - \lambda_j \mathbb{I}\} \cdot \mathbf{R}_j = O$$

or in terms of components

$$\sum_{n=1}^{N} A_{mn} X_{nj} = X_{mj} \lambda_j = O$$

The similarity transformation of $\mathbb{A}$ to the diagonal $\mathbb{\Gamma}$ can be written

$$\mathbb{S} \cdot \mathbb{A} \cdot \mathbb{S}^{-1} = \mathbb{\Gamma}$$

or multiplying through by S^{-1}

$$A \cdot S^{-1} = S^{-1} \cdot \Gamma$$

The mth equation of this set is

$$\sum_{n=1}^{N} A_{mn} S_{nj}' = \sum_{n=1}^{N} S_{mn}' \Gamma_{nj}$$

but

$$\Gamma_{nj} = \delta_{nj} \lambda_j; \text{ and}$$

$$\sum_{n=1}^{N} A_{mn} S_{nj}' = S_{mj}' \lambda_j$$

Therefore since S_{mj}' and X_{mj} satisfy the same equations,

$$S_{mj}' = X_{mj}$$

In two dimensions

$$S^{-1} = \begin{pmatrix} X_{11} & X_{12} \\ X_{21} & X_{22} \end{pmatrix}$$

Also we find that

$$\frac{X_{11}}{X_{21}} = \cot \alpha = \frac{\lambda_1 - A_{11}}{A_{12}} = \frac{A_{21}}{\lambda_1 - A_{22}}$$

In our problem concerning the ellipse, α turns out to be $-30°$.
In the problem of the hyperbola $\alpha = +45°$.
If the matrix A is symmetric

$$\text{i.e. } A = \tilde{A}$$

then it can be shown that the eigenvector solutions are mutually orthogonal (providing that two or more eigenvalues are *not* the same. An equality of the eigenvalues is called a degeneracy).

Consider the eigenvalue problem defining $\mathbf{R}_n$ and the *transposed* eigenvalue problem defining $\tilde{\mathbf{R}}_m$.

$$\tilde{\mathbf{R}}_m \cdot \Big\{ (\mathbb{A} - \lambda_n \mathbb{I}) \cdot \mathbf{R}_n = \mathrm{O}$$

$$\tilde{\mathbf{R}}_m \cdot (\widetilde{\mathbb{A} - \lambda_m \mathbb{I}}) = \mathrm{O} \Big\} \cdot \mathbf{R}_n$$

Multiply the first equation by $\tilde{\mathbf{R}}_m$ from the left; multiply the second equation by $\mathbf{R}_n$ from the right; and **subtract.**

$$\tilde{\mathbf{R}}_m \cdot \mathbb{A} \cdot \mathbf{R}_n - \tilde{\mathbf{R}}_m \cdot \tilde{\mathbb{A}} \cdot \mathbf{R}_n - (\lambda_n - \lambda_m)\, \tilde{\mathbf{R}}_m \cdot \mathbf{R}_n = \mathrm{O}$$

If $\mathbb{A} = \tilde{\mathbb{A}}$ then

$$(\lambda_n - \lambda_m)\, \tilde{\mathbf{R}}_m \cdot \mathbf{R}_n = \mathrm{O}$$

When $n \neq m$, $(\lambda_n - \lambda_m)$ is **not zero** (unless degenerate).

Therefore

$$\tilde{\mathbf{R}}_m \cdot \mathbf{R}_n = \mathrm{O};\ m \neq n$$

If the R's are normalized, i.e., unit vectors,

$$\tilde{\mathbf{R}}_m \cdot \mathbf{R}_n = \delta_{mn}$$

Summary and Applications

It will again prove convenient to summarize the steps involved in solving either a quadratic or an eigenvalue problem.

Given an equation of the type

$$\tilde{\mathbf{r}} \cdot \mathbb{A} \cdot \mathbf{r} = \mathrm{K}$$

or

$$\{\mathbb{A} - \lambda \mathbb{I}\} \cdot \mathbf{R} = \mathrm{O}$$

we proceed in the following manner:

1. The eigenvalues are obtained as roots of the algebraic equation

$$\text{Determinant}\,(\mathbb{A} - \lambda\mathbb{I}) = |\mathbb{A} - \lambda\mathbb{I}| = O$$

If $\mathbb{A}$ is an $N \times N$ square array there should be N roots λ_j.

2. The eigenvector $\mathbf{R}_j$ is obtained by substituting the specific root or eigenvalue λ_j;

$$\{\mathbb{A} - \lambda_j\mathbb{I}\} \cdot \mathbf{R}_j = O$$

or

$$\sum_{n=1}^{N} A_{mn}\, X_{nj} = X_{nj}\, \lambda_j$$

3. The orthogonal transformation $\mathbb{S}$ which takes $\mathbb{A}$ to its diagonal form $\mathbb{\Gamma}$ is obtained from the relation between the elements of $\mathbb{S}^{-1}$ and the components of $\mathbf{R}_j$.

$$S_{mj}' = X_{mj}$$

The importance of the eigenvalue problem is much greater than one would surmise by examining its applications to geometric quantities.

The problem is used in physics to solve a large variety of problems.

Many problems involving the small amplitude oscillations of coupled harmonic oscillators reduce in form to the eigenvalue problem.

An example can be indicated.

Consider two point masses suspended by three springs in a linear array between two rigid walls.

Assume that the motion is confined to a linear vertical motion for each mass.

The equations of motion for this coupled system constitute an eigenvalue problem. In this case the eigenvalues turn out to be the two characteristic frequencies of the motion. The eigenvectors of this problem describe the two characteristic motions or modes as they are called.

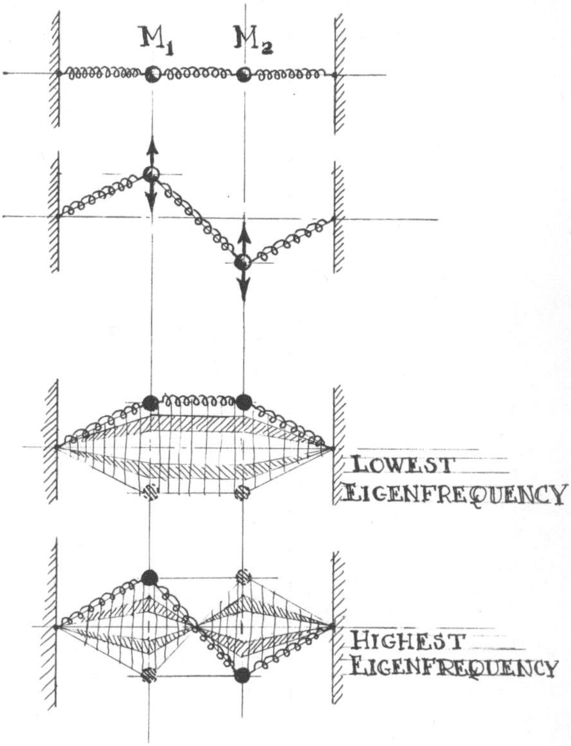

The lowest frequency mode (for equal masses) is a motion in which the two masses are in phase as shown.

The higher frequency mode consists of a motion in which the masses are out of phase; that is to say that when one has a maximum displacement *up* the other has a maximum displacement *down*.

All other motions of the system are constituted by a linear combination of the two fundamental modes.

In order to illustrate this more vividly we can set up a simple problem and solve it.

In the diagram to follow assume that the verticle displacements of M_1 and M_2 are sufficiently small in order that the tensions in the springs are essentially constant and equal to T. Also let us take the case in which the masses are equal, i.e. $M_1 = M_2 = M$.

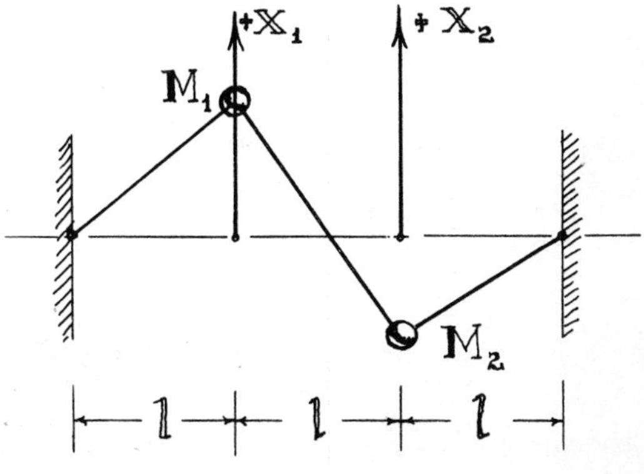

Because the motion is vertical we only consider the vertical components of T. The sum of the vertical forces on M_1 (neglect gravity) is

$$\sum F_1 = \frac{-x_1}{l} T - \frac{(x_1 - x_2)}{l} T$$

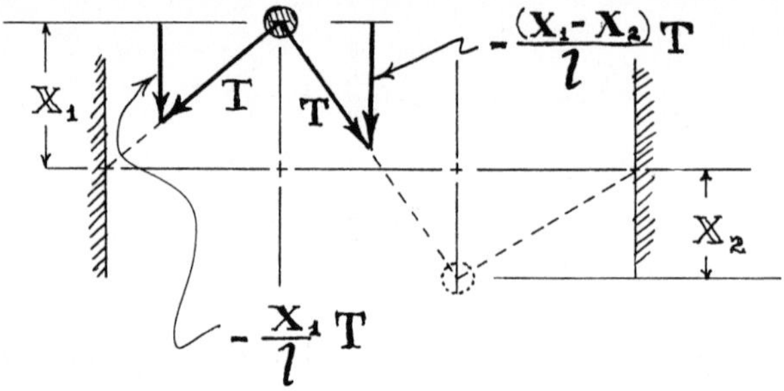

The sum of the vertical forces on M_2 is

$$\sum F_2 = \frac{-x_2}{l}T - \frac{(x_2 - x_1)}{l}T$$

By Newton's 2nd Law the vector sum of the forces on the body is equal to time rate of change of the vector momentum. In this

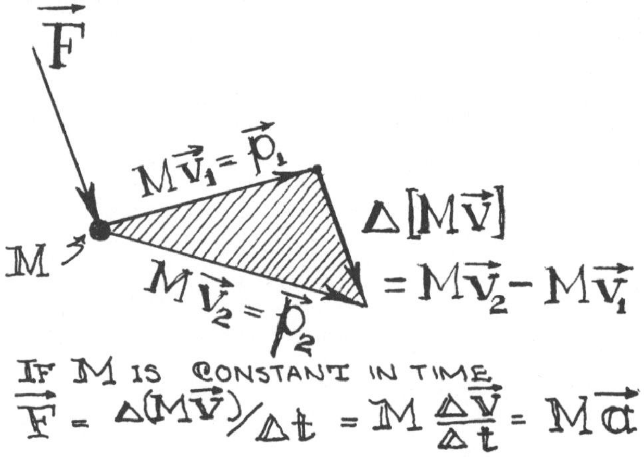

case the masses are constant in time therefore the sum of the vector forces is equal to the mass times the vector acceleration. The forces are vertical thus we need only consider the accelerations along x_1 and x_2 (shown in the diagram).

$$M\,a_1 = -\frac{2x_1}{l}T + \frac{x_2}{l}T$$

and

$$M\,a_2 = +\frac{x_1}{l}T - \frac{2x_2}{l}T$$

To solve these equations we assume that the displacements x_1 and x_2 are harmonic, i.e.

$$x_k = b_k \cos wt + c_k \sin wt$$

Under such an assumption the acceleration is proportional to the displacement. This will be proven in Chapter 5.

$$a_k = -w^2 (b_k \cos wt + c_k \sin wt) = -w^2 x_k$$

Upon substituting these values for the acceleration into Newton's 2nd Law we obtain

$$-Mw^2x_1 = \frac{-2x_1}{l}T + \frac{x_2}{l}T$$

$$-Mw^2x_2 = \frac{x_1}{l}T - \frac{-2x_2}{l}T$$

Rewriting

$$-2\Omega^2 x_1 + \Omega^2 x_2 = -w^2 x_1$$

and

$$\Omega^2 x_1 - 2\Omega^2 x_2 = -w^2 x_2$$

where

$$\Omega^2 = T/Ml$$

If we now construct a column vector $\boldsymbol{\rho} = \begin{bmatrix} x_1 \\ x_2 \end{bmatrix}$

and a matrix

$$A = \begin{pmatrix} 2\Omega^2 & -\Omega^2 \\ -\Omega^2 & 2\Omega^2 \end{pmatrix}$$

These two linear equations can be written

$$\{A - w^2 I\} \cdot \rho = O$$

A non-trivial solution exists for the equation when

$$| A - w^2 I | = O$$

Thus there are two possible values of w^2;

$$| A - w^2 I | = \begin{vmatrix} (2\Omega^2 - w^2) & -\Omega^2 \\ -\Omega^2 & (2\Omega^2 - w^2) \end{vmatrix} = O$$

or

$$(2\Omega^2 - w^2)^2 - \Omega^4 = O$$

giving the roots

$$w_1{}^2 = \Omega^2 = T/Ml$$

$$w_2{}^2 = 3\Omega^2 = 3T/Ml$$

There are two possible values of w^2, therefore the most general solution must be a linear combination of these. Let

$$x_k = \sum_{n=1}^{2} \alpha_{kn} \{b_n \cos w_n t + c_n \sin w_n t\}$$

Later it can be demonstrated that b_1, c_1, b_2 and c_2 are evaluated from the initial conditions. The initial conditions are the values of x_1 and x_2 and their velocities at time $t = 0$. Returning to the solution x_k in terms of w_1 and w_2 we can write down the acceleration;

$$a_k = -\sum_{n=1}^{2} \alpha_{kn} w_n^2 \{b_n \cos w_n t + c_n \sin w_n t\}$$

Substituting this into the equations representing Newton's 2nd Law (written in the matrix form rather than as sets of equations),

$$\mathbf{a} = \begin{bmatrix} a_1 \\ a_2 \end{bmatrix} = \mathrm{A} \cdot \boldsymbol{\rho} = \begin{pmatrix} 2\Omega^2 & -\Omega^2 \\ -\Omega^2 & 2\Omega^2 \end{pmatrix} \begin{bmatrix} x_1 \\ x_2 \end{bmatrix}$$

or

$$a_k = \sum_{l=1}^{2} A_{kl}\, x_l$$

Upon substitution for a_k (the kth acceleration) we obtain

$$-\sum_{n=1}^{2} \alpha_{kn} w_n^2 \{b_n \cos w_n t + c_n \sin w_n t\} =$$

$$\sum_{l=1}^{2} A_{kl} \sum_{n=1}^{2} \alpha_{ln} \{b_n \cos w_n t + c_n \sin w_n t\}$$

Collecting like coefficients of the bracketed terms and changing the order of summing on the right hand side,

$$\sum_{n=1}^{2} \{\sum_{l=1}^{2} A_{kl}\, \alpha_{ln} - \alpha_{kn} w_n^2\} \cdot \{b_n \cos w_n t + c_n \sin w_n t\} = O$$

For this equation to vanish

$$\sum_{l=1}^{2} A_{kl}\, \alpha_{ln} - \alpha_{kn} w_n^2 = O$$

or

$$\{A - w_n^2 \, I\} \cdot \boldsymbol{\alpha}_n = O$$

where

$$\boldsymbol{\alpha}_n = \begin{bmatrix} \alpha_{1n} \\ \\ \alpha_{2n} \end{bmatrix}$$

Thus this coupled oscillator problem reduces to the **eigenvalue problem.** Knowing that

$$w_1^2 = \Omega^2$$

and

$$w_2^2 = 3\Omega^2$$

The reader should be in a position to show that for this particular problem,

$$\alpha_1 = \frac{1}{\sqrt{2}} \begin{bmatrix} 1 \\ \\ 1 \end{bmatrix} = \begin{bmatrix} \alpha_{11} \\ \\ \alpha_{21} \end{bmatrix}$$

and

$$\alpha_2 = \frac{1}{\sqrt{2}} \begin{bmatrix} 1 \\ \\ -1 \end{bmatrix} = \begin{bmatrix} \alpha_{12} \\ \\ \alpha_{22} \end{bmatrix}$$

The general solution can be written

$$\boldsymbol{\rho} = \begin{bmatrix} x_1 \\ x_2 \end{bmatrix} = \sum_{n=1}^{2} \boldsymbol{\alpha}_n \{b_n \cos w_n t + c_n \sin w_n t\}$$

or

$$x_1 = \alpha_{11} \{b_1 \cos w_1 t + c_1 \sin w_1 t\} + \alpha_{12} \{b_2 \cos w_2 t + c_2 \sin w_2 t\}$$

Using the information that the velocities V_j are given by*

$$V = \frac{d\boldsymbol{\rho}}{dt} = \begin{bmatrix} \dfrac{dx_1}{dt} \\ \dfrac{dx_2}{dt} \end{bmatrix} = \begin{bmatrix} V_1 \\ V_2 \end{bmatrix} = \sum_{n=1}^{2} \boldsymbol{\alpha}_n w_n \{-b_n \sin w_n t + c_n \cos w_n t\}$$

The reader should be able to demonstrate from the orthogonality of the eigenvectors α_n that;

$$b_m = \tilde{\boldsymbol{\alpha}}_m \cdot \boldsymbol{\rho}\,(t = o) = x_1\,(t = o)\,\alpha_{1m} + x_2\,(t = o)\,\alpha_{2m}$$

and

$$c_m = \left(\frac{1}{w_m}\right) \alpha_m \cdot \mathbf{V}\,(t = o) = V_1\,(t = o)\,\alpha_{1m} + V_2\,(t = o)\,\alpha_{2m}$$

* The derivatives shown here will be completely discussed in Chapter 5. They are included for completeness and need not be understood in detail.

CHAPTER 4

Functions

A. *Introduction*

Tables of Relations

x	y
a_1	b_1
a_2	b_2
a_3	b_3
$\vdots$	$\vdots$
a_N	b_N

MANY TIMES IN EXPERIENCE one encounters correlated numbers. More specifically given two numbers x and y say arising from a set of observations; for every value of y there is a value of x. We can tabulate corresponding values of x and y.

Suppose for instance we observe N different values of x and obtain a value of y for each x. A typical example of such a procedure would be to write down the temperature in a room as a function of the height from the floor at a specified position on the floor. Thus for every vertical distance h from the floor there is a corresponding temperature T (h).

The table of corresponding values of x and y can be plotted. Such a plot is shown below.

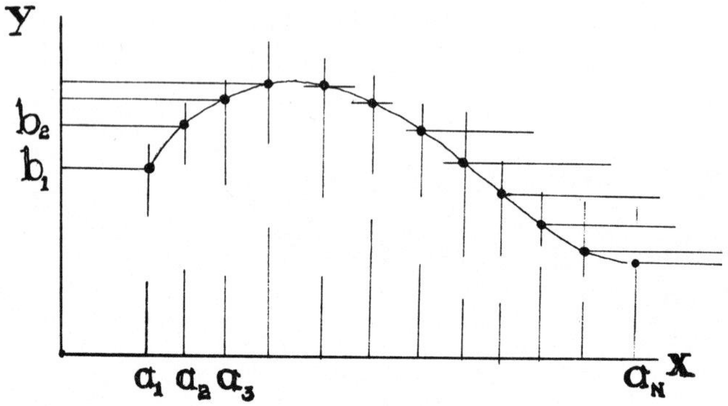

A smooth curve has been drawn through the points. With a discreet set of points the curve which one draws through points and connecting them is *not unique.*

To illustrate this idea two other curves are shown which satisfy equally the correspondence exhibited in the table of Section A.

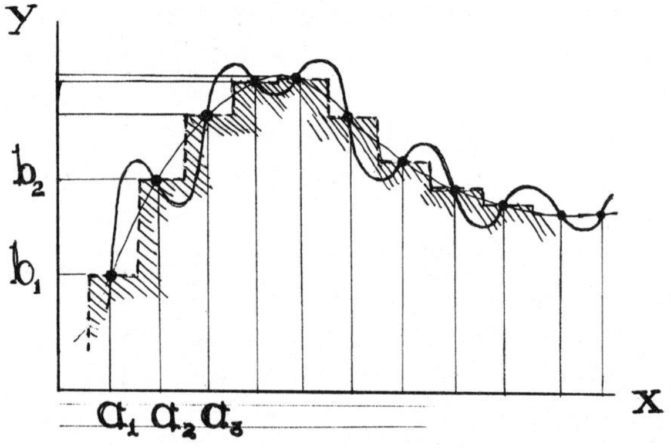

We then realize that in order to draw a unique curve through a discreet set of points a detailed specification of the behavior of the curve in the interval between points is necessary.

B. The Definition of a Function

If we specify an algebraic relation between the variables x and y, the variable y can be defined as a function of the independent variable x.

f(x) is said to be a function of the independent variable x in a specified interval x = a to x = b if there are **one** or **more values** of f(x) for each value of x in the interval.

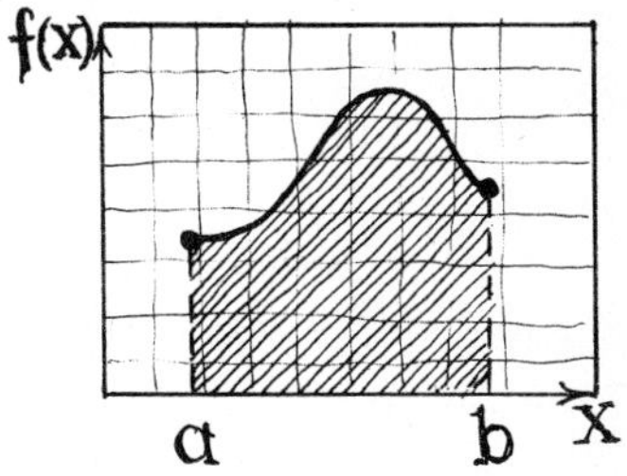

C. Definition of an Interval

The interval is said to be **closed** if the end points a and b are included.

Closed Interval; $a \leq x \leq b$. The interval is said to be open if the function f(x) is defined for all points in the interval between a and b but excluding the end points a and b.

Open Interval; $a < x < b$.

Interval Open at one end; $a \leq x < b$ or $a < x \leq b$. To illustrate the types of physical situations in which the **interval of definition** must be clearly specified, consider a length of string L vibrating between two rigid thin plates which support it. For the sake of simplicity assume that the string vibrates as a pure sine wave; the y displacement at any point x being given by

$$y(x, t) = a_0 \sin\left(\frac{\pi}{L} x\right) \sin wt$$

Note carefully that this equation must be qualified as holding in the region $o \leq x \leq L$.

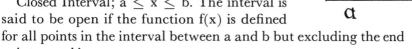

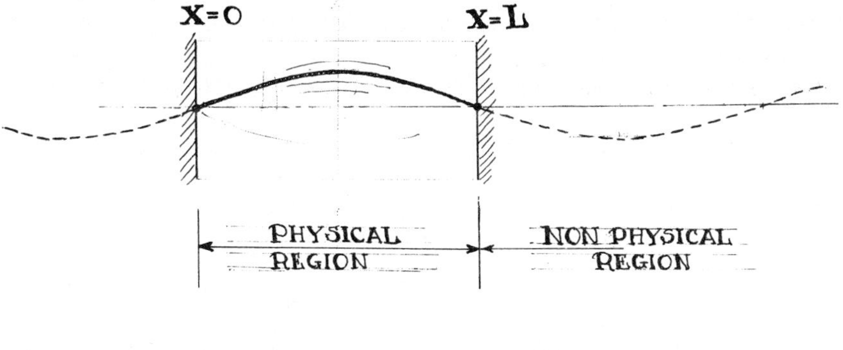

Because the physical problem states that the string is present only from $x = 0$ to $x = L$ it is not proper physically to consider the solution outside of the interval.

This example is characteristic of many problems in which the model function exists mathematically outside the region of definition but has no physical meaning. There are interesting mathematical cases in which the function may not be defined at a point or in a region. In such cases care must be exercised in using the function because spurious results may arise from the inadvertent inclusion of a point at which the function is not defined.

As an example suppose we look at the area enclosed between the x axis and the curve given by $1/x^2$ in the interval from a to b (a and b both positive). If by chance we attempt to let "a" go to $x = 0$ the area under curve is undefined.

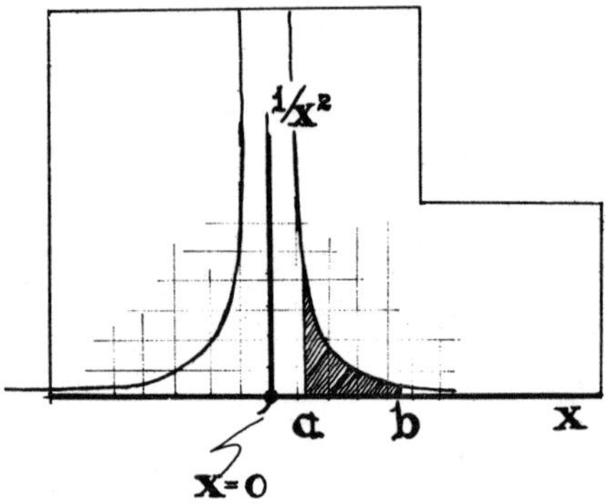

The area in the interval $a \leq x \leq b = \frac{1}{a} - \frac{1}{b}$

From the relation above we see that the term $\frac{1}{a}$ is undefined at $a = 0$. This example also demonstrates that the interval cannot be taken in such a manner as to include the point $x = 0$. In other words we cannot discuss the area under the curve when a lies on the left of $x = 0$ and when b lies on the right even though the relation $\frac{1}{a} - \frac{1}{b}$ exists.

D. Multiplicity:

For a given value of the independent variable x there can be one or more values of f(x). If there is but one value of f(x) for every value of x the function is said to be single valued. Higher multiplicities are labeled similarly.

Single Valued Functions

If for every x in the interval a → b there is **one** and only **one** value of f(x), f(x) is said to be single valued.

As an example the parabola is shown below.

$$y = 2x^2$$

or

$$f(x) = 2x^2$$

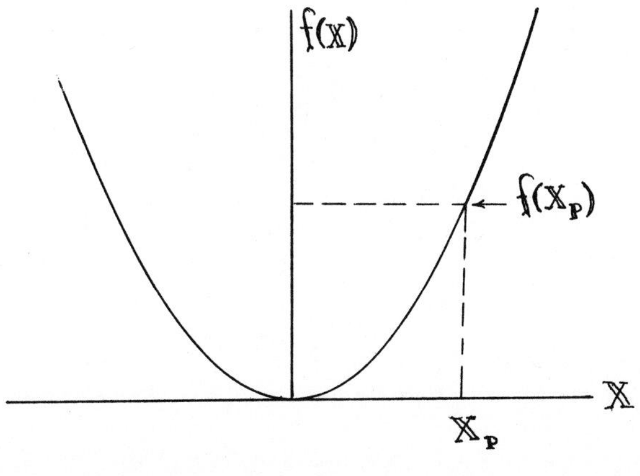

Double Valued Functions

If for every $a < x < b$ there are **two** and **only two values** of the function, the function is said to be *Double Valued* in that interval.

To illustrate such a function we consider the inverse function

$$x = \pm \sqrt{y/2}$$

$$\text{or } g(y) = \pm \sqrt{\frac{y}{2}}$$

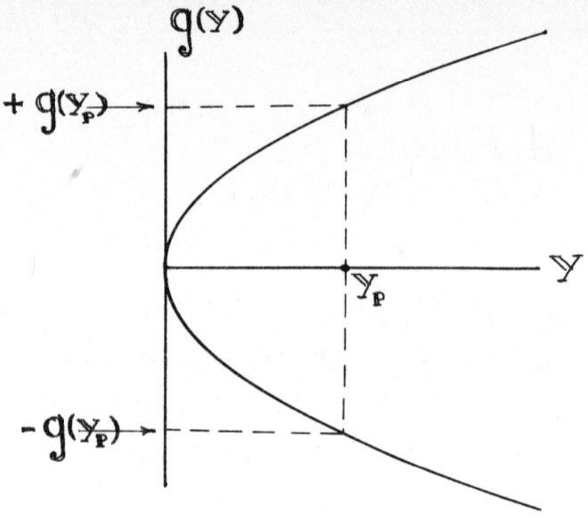

It should be noticed in the case above that the function g(y) is **double valued** in an interval **open** at one end

$$g(y) \text{ double valued } 0 < y \leq N$$

where N can be taken to be any arbitrarily large number.

$$g(y) \text{ is single valued at } y = 0$$

A **second example** is the circle

$$x^2 + y^2 = R^2$$

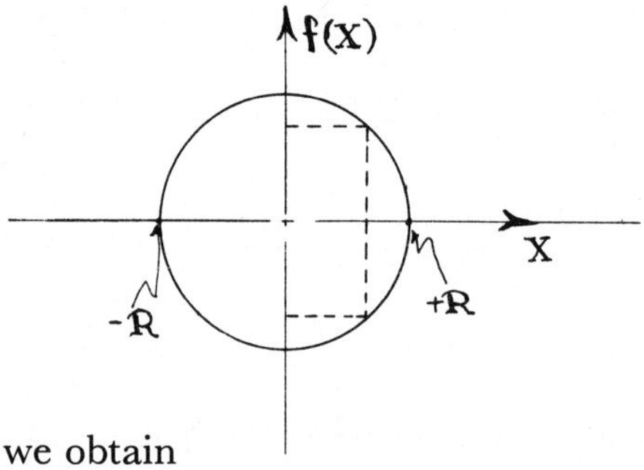

Solving for y we obtain

$$f(x) = y = \pm \sqrt{R^2 - x^2}$$

If we require that f(x) be a real number then the reader can easily verify that

$f(x)$ is double valued $-R < x < R$

$f(x)$ is single valued at $x = \pm R$

A Function which is Triple Valued in the Interval

$$-b < x < b$$

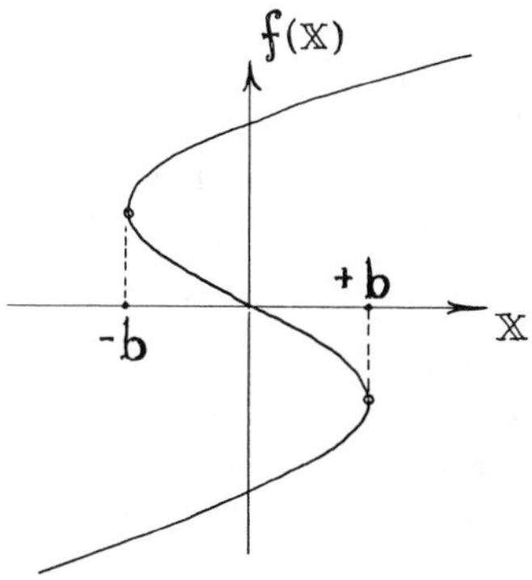

is illustrated at the right. This particular function is **double valued** at

$$x = \pm b$$

and is **single valued** in the two open ended intervals

$$-N \leq x < -b \text{ and } b < x \leq N$$

Higher Multiplicity

There are many examples of functions which have a higher multiplicity. One familiar example is the inverse trigonometric function. The sine and cosine functions are single valued, i.e.

$$f(x) = \sin x \text{ is single valued.}$$

The inverse function

$$g(y) = \sin^{-1}(y) \qquad -1 \leq y \leq +1$$

is multivalued. The interval is interesting in that the multiplicity in the open interval

$$-1 < y < 1$$

is twice the multiplicity at the ends of the interval.

$$|y| = 1$$

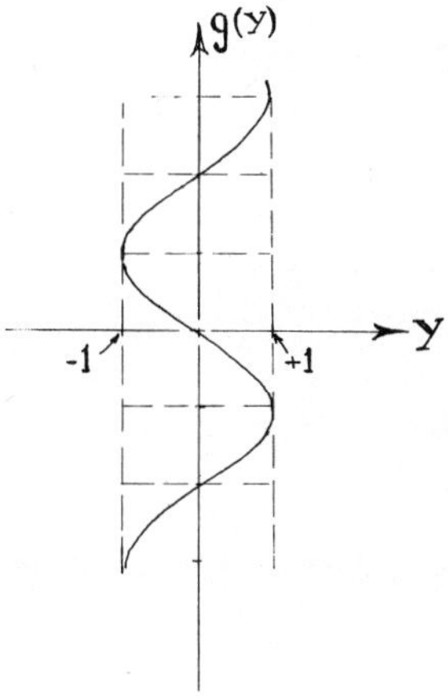

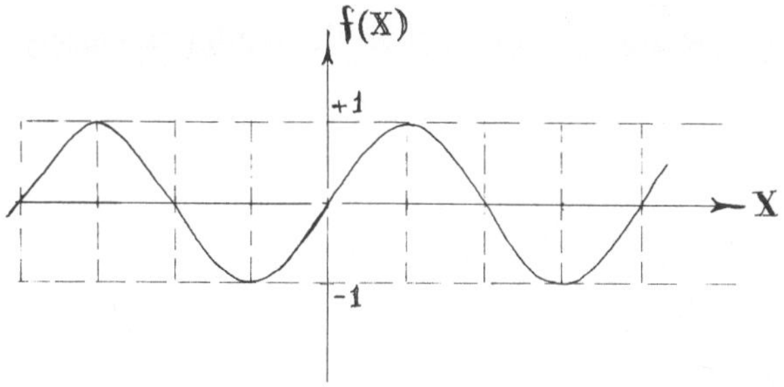

E. The Slope at a Point

Definition

Consider the graph of a curve f(x).

The tangent to the curve f(x) at a specified point P is a straight line which we shall label L. This line L makes an angle α_p with respect to the x axis.

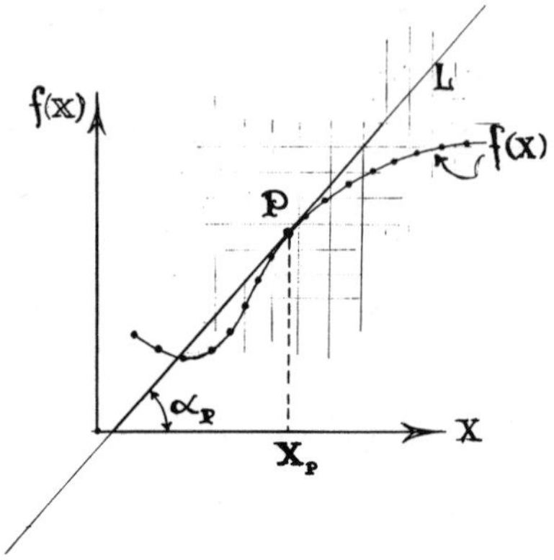

The slope at P = tan α_p.

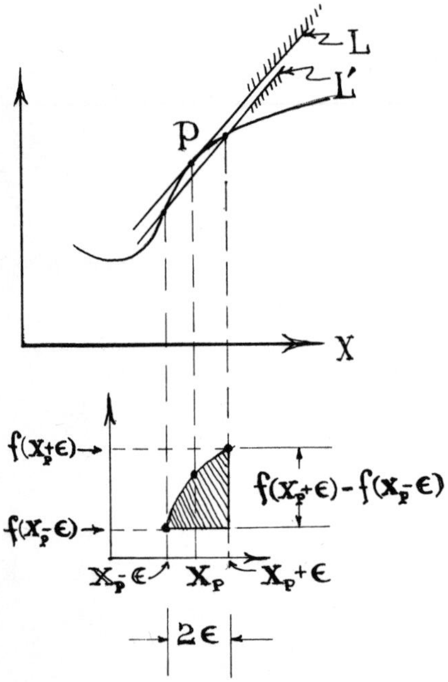

A line tangent to a curve f(x) at **one** point touches the curve once and only once in the vicinity of the point. Because the tangent line L as we have drawn it is extended to the x axis the curve may oscillate and cross this line L at other points. Such a situation of course has no bearing on our argument.

For this reason we will deal with the angle between L and a line parallel to the x axis in the vicinity of P.

In a slope measuring operation, performed with a straight edge, it is difficult to judge the exact slope. There are simple algebraic methods for **approximating** the slope at a point.

Later we shall see that these algebraic methods lead to a technique by which we determine the **exact slope** at P. These later methods are those of the **differential calculus.**

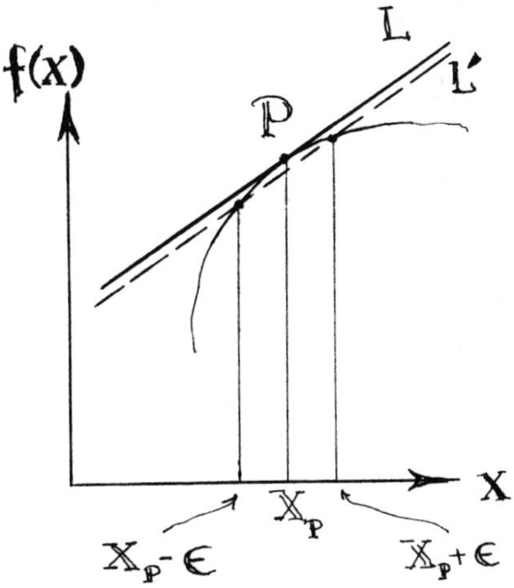

The tangent at a point can be obtained approximately by considering a **small segment** of **arc** at P. At P the tangent touches the curve f(x) at **one** point in a region very close to the point in question.

We observe from the diagram at the right that a line L′ passing through the points $(x_p + \epsilon)$ and $(x_p - \epsilon)$ at the **ends** of the **arc** is approximately **parallel** to L.

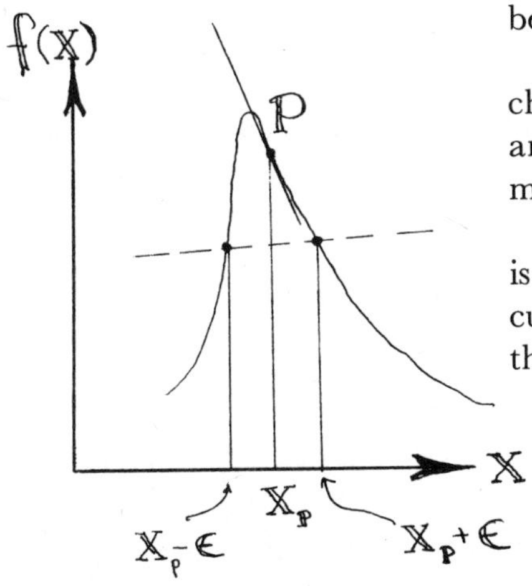

Later we shall find that the derivative is obtained by allowing ϵ to become arbitrarily small.

If 2ϵ is of the same order as the change in f(x) between $(x_p + \epsilon)$ and $(x_p - \epsilon)$ the approximation may be relatively good.

On the left an extreme example is presented of a case in which the curve fluctuates to a great extent in the interval chosen.

The line L makes an angle α_p with respect to the x axis (or a line parallel to the x axis). The line L′ makes an angle α_p' with the x axis (or a line parallel to the x axis).

As the length of **arc** (or 2ϵ) becomes smaller the value of the angle α_p' approaches the value of the angle α_p. We assume that as ϵ becomes arbitrarily small or zero; α_p' becomes equal to α_p except for certain cases which we shall discuss. If the slope is undefined at P then our arguments as presented above do not hold.

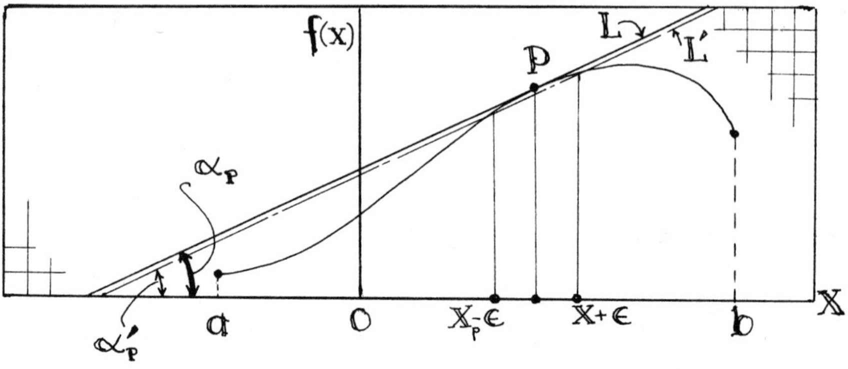

The slope of f(x) at P is defined as,

$$\text{slope at P} = \tan \alpha_p$$

It is apparent that the slope can be **approximated** by computing $\alpha_p{}'$ utilizing two points $x_p + \epsilon$ and $x_p - \epsilon$ close to P.

$$\text{slope at P} \simeq \tan \alpha_p{}'$$

We can compute the approximate slope from the values of f(x) at the two points $x_p + \epsilon$ and $x_p - \epsilon$.

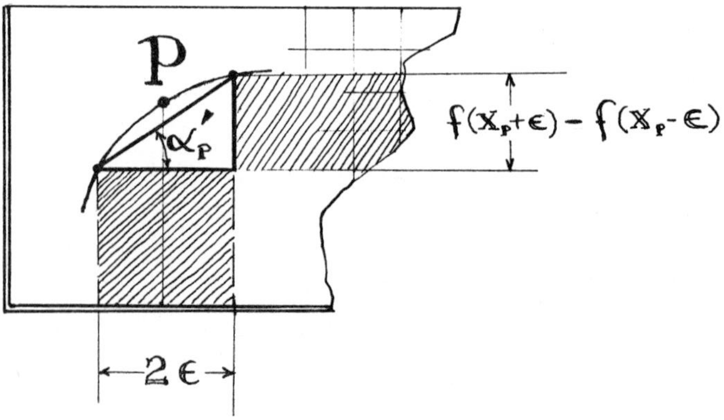

From the diagram we see that

$$\tan \alpha_p' = \frac{f(x_p + \epsilon) - f(x_p - \epsilon)}{2\epsilon}$$

As an example of this type of approximation consider the parabola which we used as an example earlier.

$$f(x) = 2x^2$$

Let us compute the slope at $x = 1$ (this is x_p) using as the interval $\epsilon = 1/4$. Thus $x_p + \epsilon = 5/4$ and $x_p - \epsilon = 3/4$.

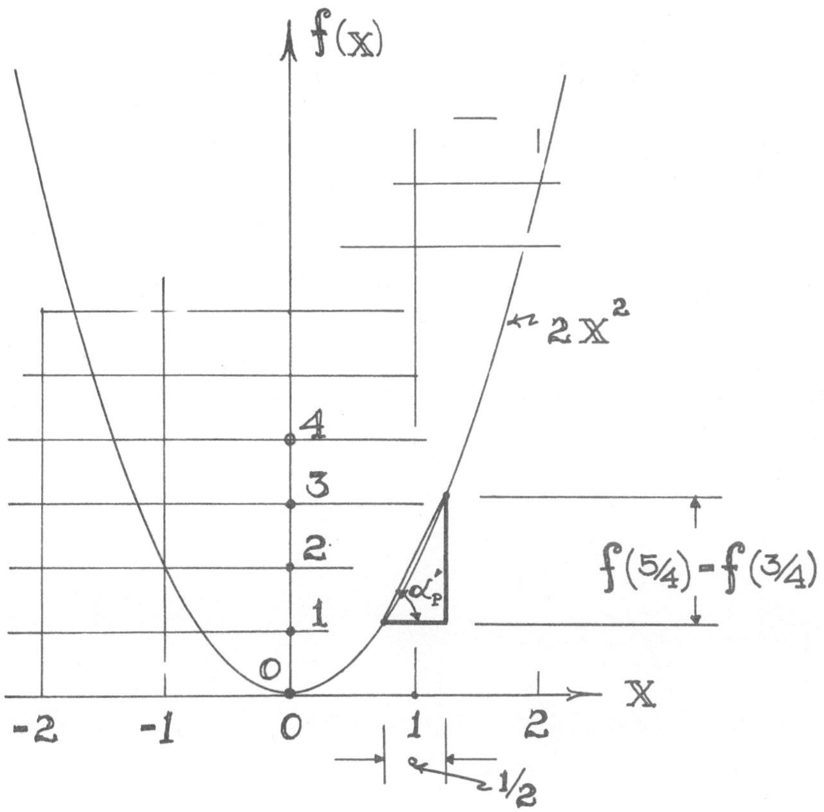

Then

$$\tan \alpha_p' = \frac{f(5/4) - f(3/4)}{2 \times 1/4}$$

$$= \frac{\frac{25}{8} - \frac{9}{8}}{1/2} = 4 = \text{slope } x = 1$$

This is a fortuitous result since the true slope at $x = 1$ is actually 4.

It will be of interest in the sections to follow to note here that one can now plot the slope as a function of x. Since for every x we have a $\tan \alpha_x$, we can plot $\tan \alpha_x$ *vs* x. In the diagrams to follow the simple example of the parabola is used to illustrate the plot of $\tan \alpha_x$ as a function of x.

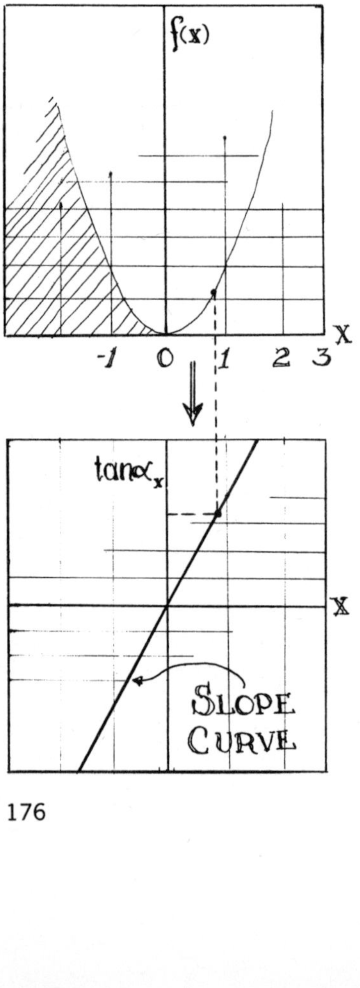

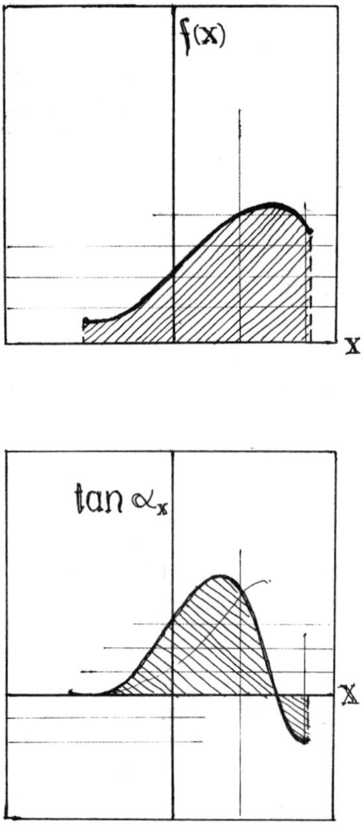

It is interesting in connection with the discussion of the slope to find that if we plot the displacement of a body as a function of time, the slope of the displacement vs time graph is the magnitude of the velocity in direction of the displacement (the magnitude of a velocity is called the speed).

$$|\mathbf{V}_y|_p = \text{slope of } y(t) \text{ at } t_p$$

If we now plot $|\mathbf{V}_y|$ the slope of y as a function of t, the slope of the $|\mathbf{V}_y|$ curve at any point is a measure of the magnitude $|\mathbf{a}_y|$ of the acceleration in the direction of the displacement.

$$|\mathbf{a}_y|_{tp} = \text{slope of } |\mathbf{V}_y| \text{ at } t_p$$

Thus the magnitude of the acceleration is the second slope of the displacement curve.

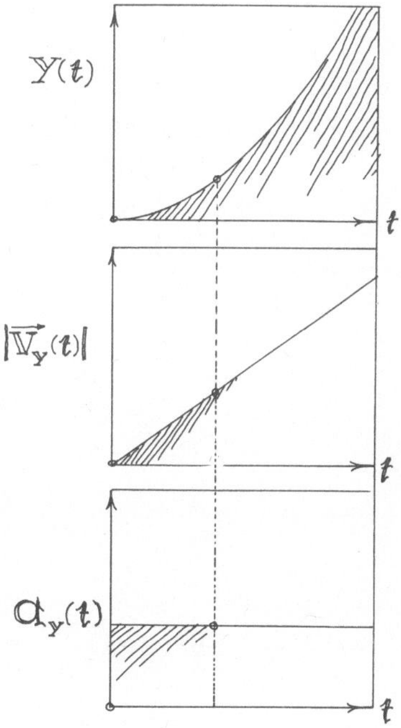

1.) Take a curve given by f(x) in the interval $a \leq x \leq b$. To make the representation simple shift the origin to coincide with a.
2.) Divide the interval $a \rightarrow b$ into N sub-intervals of equal width ϵ.
3.) The values of f(x) for every interval can be listed in a column i.e.,

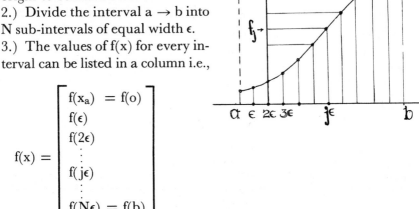

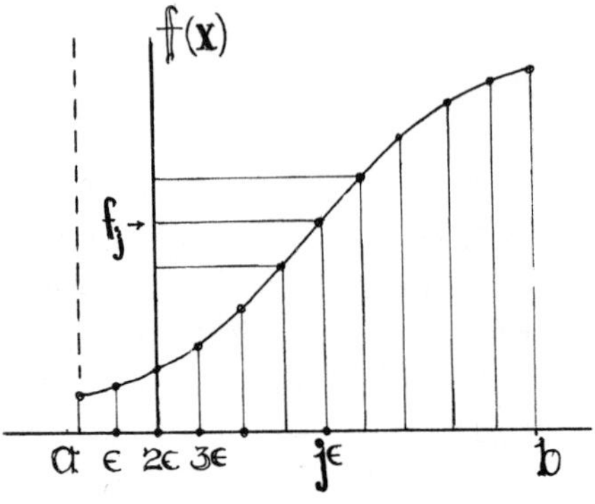

$$f(x) = \begin{bmatrix} f(x_a) = f(o) \\ f(\epsilon) \\ f(2\epsilon) \\ \vdots \\ f(j\epsilon) \\ \vdots \\ f(N\epsilon) = f(b) \end{bmatrix}$$

This tabulation of the values of f(x) can be thought of as a **column matrix** of N + 1 **dimensions.**

When this tabulation is used the function can be written using the following subscript notation,

$$f(x) = \begin{bmatrix} f(o) \\ f(\epsilon) \\ \vdots \\ f(j\epsilon) \\ \vdots \\ f(N\epsilon) \end{bmatrix} = \begin{bmatrix} f_o \\ f_1 \\ \vdots \\ f_j \\ \vdots \\ f_N \end{bmatrix}$$

4.) The computation of the **approximate** slope at a point $x = j\epsilon$ (where j is an integer less than N) is a **linear combination** of the values of f(x) evaluated at $(j - 1)\epsilon$ and $(j + 1)\epsilon$.

If $f(x)$ = the function at x

let

$d(x)$ = the slope at x

Thus

$$d(j\epsilon) = \frac{f([j+1]\epsilon) - f([j-1]\epsilon)}{2\epsilon} = \frac{f_{j+1} - f_{j-1}}{2\epsilon}$$

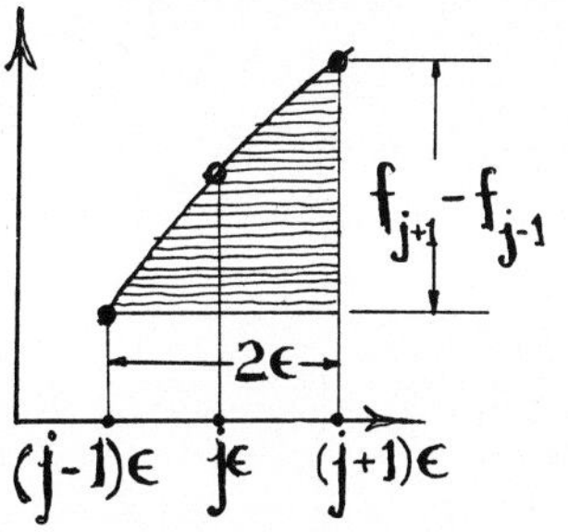

Because $d(x)$ is a linear transformation of $f(x)$ we can in general write

$$d(m\epsilon) = D_{mo}\, f_o + D_{m1}\, f_1 + D_{m2}\, f_2 + \ldots\ldots D_{mN}\, f_N = \sum_{n=O}^{N} D_{mn}\, f_n$$

In order to make the notation consistent with the customary matrix notation, write

$$f(n\epsilon) = f_n$$
$$d(m\epsilon) = d_m$$

Then

$$d_m = \sum_{n=O}^{N} D_{mn}\, f_n$$

By examining the slope at any point $x = j\epsilon$ i.e. $d(j\epsilon) = d_j$ we find that

$$D_{mn} = O \qquad \text{for } n < (m-1)$$
$$\text{for } n = m$$
$$\text{for } (m+1) < n$$

and

$$D_{m,m-1} = -\frac{1}{2\epsilon} \qquad n = m - 1$$

$$D_{m,m+1} = +\frac{1}{2\epsilon} \qquad n = m + 1$$

The slope taking operation now takes the general operator form

$$\mathbf{d} = \begin{bmatrix} d_o \\ \vdots \\ d_j \\ \vdots \\ d_N \end{bmatrix} = \mathbb{D} \cdot \mathbf{f} = \mathbb{D} \cdot \begin{bmatrix} f_o \\ \vdots \\ f_j \\ \vdots \\ f_N \end{bmatrix}$$

where

$$\mathbb{D} = \frac{1}{2\epsilon} \begin{pmatrix} 0 & 1 & 0 & 0 & 0 & 0 \\ -1 & 0 & 1 & 0 & 0 & 0 \\ 0 & -1 & 0 & 1 & 0 & 0 \\ 0 & 0 & -1 & 0 & 1 & 0 \\ 0 & 0 & 0 & -1 & 0 & 1 \\ 0 & 0 & 0 & 0 & -1 & 0 \end{pmatrix}$$

The example given utilizes a case in which $N = 6$ or in which $\mathbb{D}$ is a 6 x 6 matrix. This has been done merely to illustrate the general matrix form. Any N x N example serves equally well.

There is one inherent difficulty with the operator $\mathbb{D}$. The approximation of the slope at every point is quite good,

except** at the **ends** of the **interval.

In other words the values of d(o) and of d($N\epsilon$) are very inaccurate.

To maintain the symmetry of this matrix we shall consistently ignore all end point values of the resulting column matrices, i.e., neglect results for $x = o$ and for $x = N\epsilon$.

The operator (or matrix) $\mathbb{D}$ has many intriguing properties. The slope of the slope curve which we will label as a vector $d^{(2)}$ is the result of using $\mathbb{D}$ twice.

$$\mathbf{d}^{(2)} = \mathbb{D} \cdot \mathbf{d} = \mathbb{D} \cdot [\mathbb{D} \cdot \mathbf{f}] = \mathbb{D} : \mathbb{D} \cdot \mathbf{f}$$

The matrix $\mathbb{D} \cdot \mathbb{D}$ has a most interesting form. It measures the approximate **curvature** of **f** at a point. The reader can develop the matrix $\mathbb{D} \cdot \mathbb{D}$ by multiplication.

$$\mathbb{D}\cdot\mathbb{D} = \frac{1}{4\epsilon^2}\begin{pmatrix} 0 & 1 & 0 & 0 & 0 & \cdots \\ -1 & 0 & 1 & 0 & & \vdots \\ 0 & -1 & 0 & & & \vdots \\ 0 & & & & & \vdots \\ 0 & & & & & \vdots \\ 0 & \cdots & \cdots & \cdots & \cdots & \vdots \end{pmatrix}\begin{pmatrix} 0 & 1 & 0 & 0 & 0 & \cdots \\ -1 & 0 & 1 & 0 & & \vdots \\ 0 & -1 & 0 & & & \vdots \\ 0 & & & & & \vdots \\ 0 & & & & & \vdots \\ 0 & \cdots & \cdots & \cdots & \cdots & \vdots \end{pmatrix}$$

giving

$$\mathbb{D}\cdot\mathbb{D} = \frac{1}{4\epsilon^2}\begin{pmatrix} -2 & 0 & 1 & 0 & 0 & 0 \\ 0 & -2 & 0 & 1 & 0 & 0 \\ 1 & 0 & -2 & 0 & 1 & 0 \\ 0 & 1 & 0 & -2 & 0 & 1 \\ 0 & 0 & 1 & 0 & -2 & 0 \\ 0 & 0 & 0 & 1 & 0 & -2 \end{pmatrix}$$

By examination of the matrix above we find that at the point $x = n\epsilon$

$$(\mathbb{D}\cdot\mathbb{D}\cdot\mathbf{f})_{n\epsilon} = \frac{f_{n-2} - 2f_n + f_{n+2}}{4\epsilon^2}$$

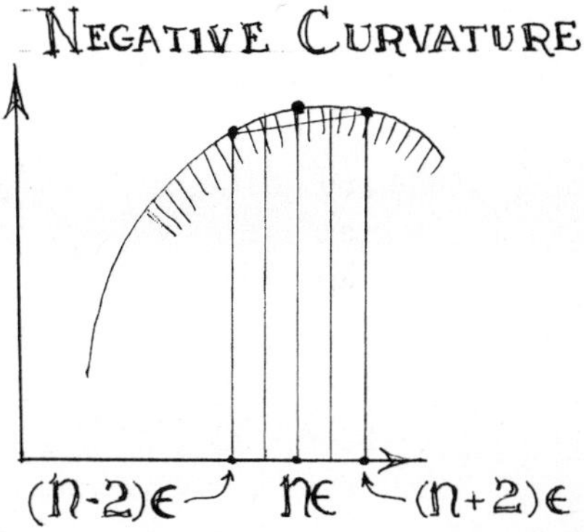

If we relate this algebraic equation to the graph shown at the right it is apparent that the operation $\mathbb{D}\cdot\mathbb{D}$ merely subtracts twice the value of f(x) evaluated at $x = n\epsilon$ from the **sum** of the values of f(x) taken at points on either side of $n\epsilon$; in other words at $(n-2)\epsilon$ and at $(n+2)\epsilon$.

If the curve f(x) is **concave** *downward* $\mathbb{D}\cdot\mathbb{D}\cdot\mathbf{f}$ is **negative** indicating a negative curvature.

If the curve is **concave** *upward* the operation $\mathbb{D}\cdot\mathbb{D}\cdot\mathbf{f}$ at the point in question is **positive,** or the curvature of **f** is positive at that point.

Because of the symmetry about the diagonal element in the matrix $\mathbb{D}$ the matrix $\mathbb{D} \cdot \mathbb{D} = \mathbb{D}^2$ appears to have some unnecessary spacing. The matrix

$$\mathbb{D}^2 = \frac{1}{4\epsilon^2}\begin{pmatrix} -2 & 1 & 0 & 0 & 0 & 0 \\ 1 & -2 & 1 & 0 & 0 & \\ 0 & 1 & -2 & 1 & 0 & \\ 0 & 0 & 1 & -2 & & \\ 0 & & & & & \\ 0 & & & & & \\ 0 & & & & & \end{pmatrix}$$

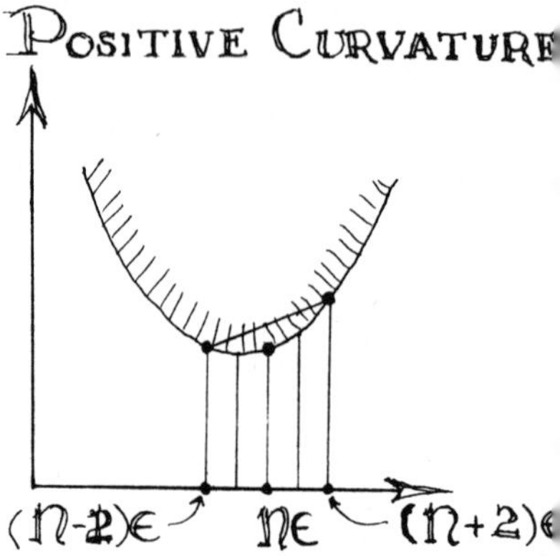

will also measure the curvature. This elimination of zeros is equivalent to removing every other element in the column matrix **f.** However for a consistent set of operations the symmetry about the diagonal should be maintained.

Finally the matrix $\mathbb{D}$ can be shown to have an inverse. In other words given d(x) can we find f(x),

$$\mathbb{D}^{-1} \cdot \mathbf{d}(x) = \mathbb{D}^{-1} \cdot \mathbb{D} \cdot \mathbf{f} = \mathbf{f}(x)$$

Later in this chapter it will be demonstrated that the inverse of $\mathbb{D}$ is the operation corresponding to taking the **area under a curve.**

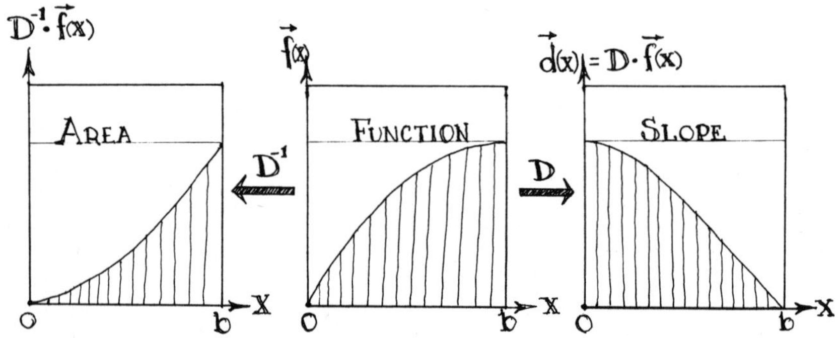

Thus it will be of some importance to compute $\mathbb{D}^{-1}$.

To perform this inverse operation, as we have indicated, the inaccuracies at the end points begin to play a very important role. One technique by which we can avoid this difficulty is that of taking an extremely large number of **sub-intervals** and work only with the central part of the matrix constantly ignoring the end points.

Another method is developed by disregarding the symmetry of $\mathbb{D}$ and settling for a slightly more inaccurate estimate of the slope. This inaccuracy can again be minimized by taking a large number of sub-intervals.

We will eliminate every other element in the vector starting with the first. In other words we eliminate the even values of j.

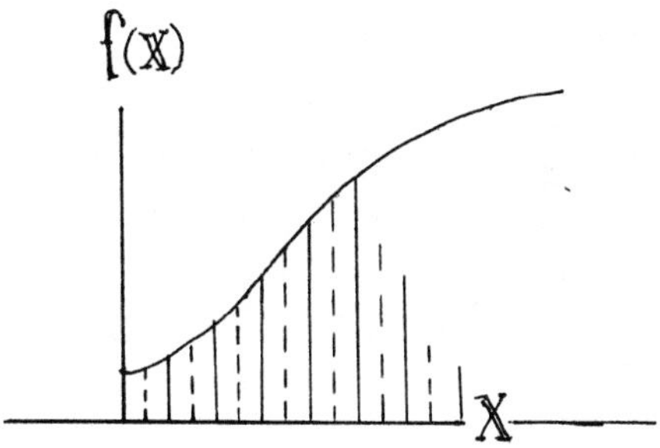

We then take a large number of intervals M and define the slope at $x = m\epsilon$ as

$$d(m\epsilon) = \frac{f(m\epsilon) - f([m-1]\epsilon)}{\epsilon}$$

Then

$$\mathbb{D} = \frac{1}{\epsilon}\begin{pmatrix} 1 & 0 & 0 & 0 & 0 & 0 \\ -1 & 1 & 0 & 0 & 0 & 0 \\ 0 & -1 & 1 & 0 & 0 & 0 \\ 0 & 0 & -1 & 1 & 0 & 0 \\ 0 & 0 & 0 & -1 & 1 & 0 \\ 0 & 0 & 0 & 0 & -1 & 1 \end{pmatrix}$$

where $\quad D_{mn} = 0 \qquad m < n < (m-1)$

$D_{mm} = +\,1/\epsilon$

and $\quad D_{m,m-1} = -\,1/\epsilon$

The operator $\mathbb{D}^2$ is much the same. However the computation is shifted slightly from the starting point.

$$\mathbb{D}^2 = \mathbb{D} \cdot \mathbb{D} = \frac{1}{\epsilon^2} \begin{pmatrix} 1 & O & O & O & O & O \\ -2 & 1 & O & O & O & O \\ 1 & -2 & 1 & O & O & O \\ O & 1 & -2 & 1 & O & O \\ O & O & 1 & -2 & 1 & O \\ O & O & O & 1 & -2 & 1 \end{pmatrix}$$

$\mathbb{D}^2$ is obtained in its more compact form. However the curvature appearing in the jth position corresponds properly to the $j - 1$ position. The operations shift 1/2 of an interval for every application. This occurs because this particular form of $\mathbb{D}$ is **shifted by one-half an interval.**

This form however is very useful for the purposes of computing $\mathbb{D}^{-1}$. Remember

$$D_{mn}' = (-1)^{m+n} \frac{\text{cof } D_{nm}}{|\mathbb{D}|}$$

Using this equation to obtain the elements of $\mathbb{D}^{-1}$ we find that

$$\mathbb{D}^{-1} = \epsilon \begin{pmatrix} 1 & O & O & O & O & O \\ 1 & 1 & O & O & O & O \\ 1 & 1 & 1 & O & O & O \\ 1 & 1 & 1 & 1 & O & O \\ 1 & 1 & 1 & 1 & 1 & O \\ 1 & 1 & 1 & 1 & 1 & 1 \end{pmatrix}$$

or

$$D_{mn}' = \epsilon \qquad \text{if } n \leq m$$
$$D_{mn}' = O \qquad \text{if } m < n$$

The suspicious reader can immediately check this form by carrying out the multiplication of

$$\mathbb{D}^{-1} \cdot \mathbb{D} = \mathbb{I}$$

$$\mathbb{D}^{-1} \cdot \mathbb{D} = \frac{\epsilon}{\epsilon} \begin{pmatrix} 1 & 0 & 0 & 0 & 0 & 0 \\ 1 & 1 & 0 & 0 & 0 & 0 \\ 1 & 1 & 1 & 0 & 0 & 0 \\ 1 & 1 & 1 & 1 & 0 & 0 \\ 1 & 1 & 1 & 1 & 1 & 0 \\ 1 & 1 & 1 & 1 & 1 & 1 \end{pmatrix}.$$

$$\begin{pmatrix} 1 & 0 & 0 & 0 & 0 & 0 \\ -1 & 1 & 0 & 0 & 0 & 0 \\ 0 & -1 & 1 & 0 & 0 & 0 \\ 0 & 0 & -1 & 1 & 0 & 0 \\ 0 & 0 & 0 & -1 & 1 & 0 \\ 0 & 0 & 0 & 0 & -1 & 1 \end{pmatrix} = \begin{pmatrix} 1 & 0 & 0 & 0 & 0 & 0 \\ 0 & 1 & 0 & 0 & 0 & 0 \\ 0 & 0 & 1 & 0 & 0 & 0 \\ 0 & 0 & 0 & 1 & 0 & 0 \\ 0 & 0 & 0 & 0 & 1 & 0 \\ 0 & 0 & 0 & 0 & 0 & 1 \end{pmatrix}$$

The matrix $\mathbb{D}^{-1}$ is called the **triangular matrix** and represents an integration or the operation of taking an area under a curve.

These operations at first sight may appear to be an unnecessary sophistication, however many problems in practical scientific work must be worked out numerically on **computers.**

The methods which we are now developing in terms of the **matrix symbolism** are a concise formulation of the types of operations which can be done on a computer.

In computer calculations the information used is invariably in tabular form. The manipulation of this information to obtain the desired result is in many cases nothing more than the manipulations which we indicate symbolically in terms of our matrices.

F. Continuity

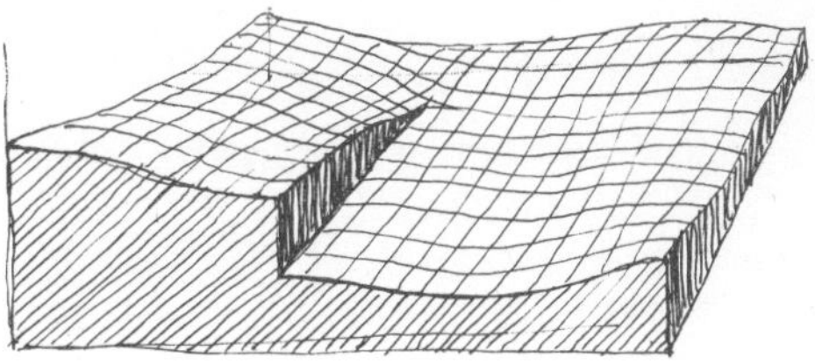

Problems of continuity in functions reach all degrees and shades of complexity. At this point we can at best merely note a few characteristic examples of continuous and discontinuous functions. We shall make a few definitions which are suited to our purpose of cataloguing the various types of curves which may be encountered.

1. Continuous Functions

We shall say that a function is continuous in the vicinity of a point P if the curve of the slope d(x) in this region has no discontinuities.

The slope curve can possess a **"cusp"** (a **kink,** or break) and yet correspond to a continuous function. We shall gain some insight concerning this last statement when we discuss the area under the curve representing the function.

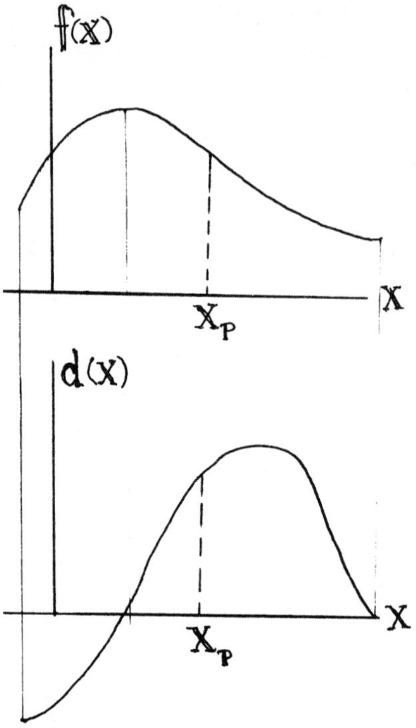

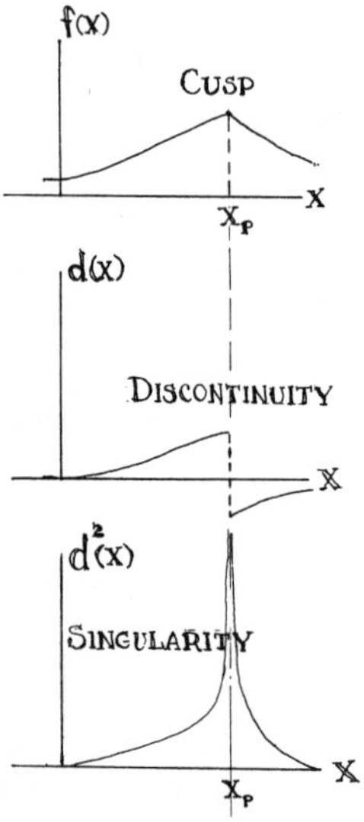

A function f(x) has a "cusp" at a point P if the slope plot is **discontinuous** at P. That is to say that the slope changes from one value to a different value at the point P, depending upon the side from which the point is approached.

If the *slope* of the d(x) curve is $d^{(2)}(x)$ at every point x, it is interesting to note that $d^{(2)}(x)$ is **singular** at P, when P is a point of discontinuity in d(x).

If the magnitude of the function becomes very large on either side of P and arbitrarily close to P, and if the function has the value N at P where N is so large that it is undefined, we say that f(x) is singular at P.

It is worth while to mention several singular functions in order to illustrate this case.

The slope of a **step function** is singular at the step.

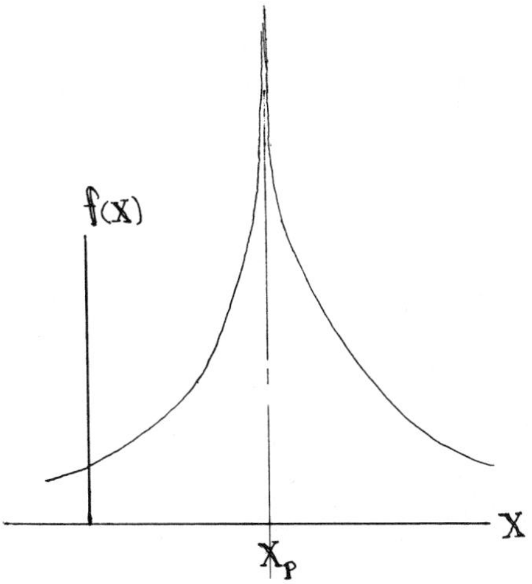

Let

$$f(x) = 0 \qquad -N \leq x < P$$

and

$$f(x) = +1 \qquad P < x \leq N$$

Note that we use an open interval to the left and a closed interval to the right in order to **define** the value of f(x) at the discontinuity P. Then the slope curve is illustrated at the right.

The function is said to be **undefined** at a singularity.

We can see how the singular slope of the step function is obtained in the limit by examining a function f′(x) which is not quite a step function.

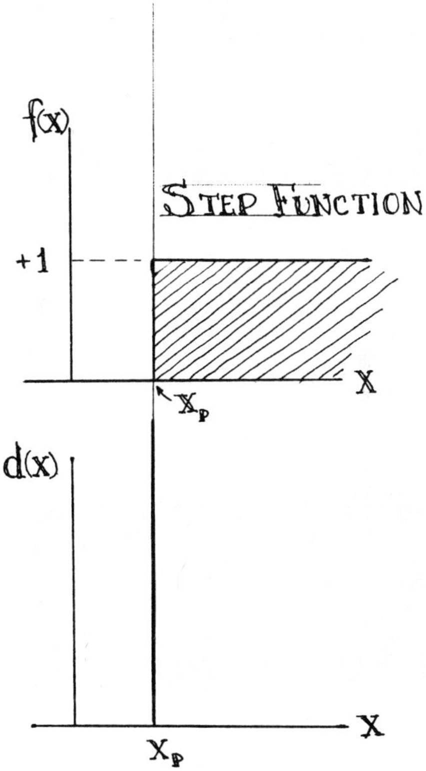

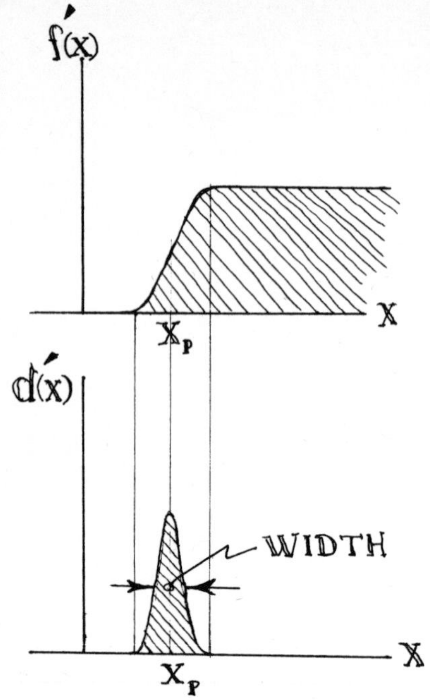

The d′(x) curve has a width W at 1/2 the **maximum value.** It should be intuitively (?) obvious that as f′(x) approaches a step function, the width W approaches **zero,** and the maximum value of d′(x) becomes infinite.

Singularities can be of many types; however, some characteristic examples are

1. The function which is positive on both sides of the singularity.

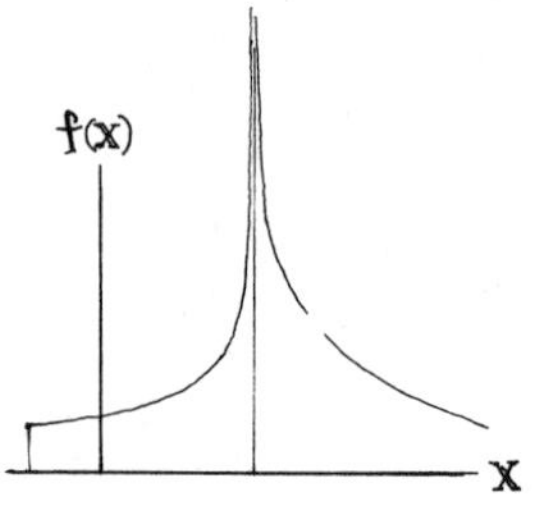

2. The function which changes sign upon passing through the singularity.

A familiar example of this is the tangent θ curve at $\theta = \pi/2$.

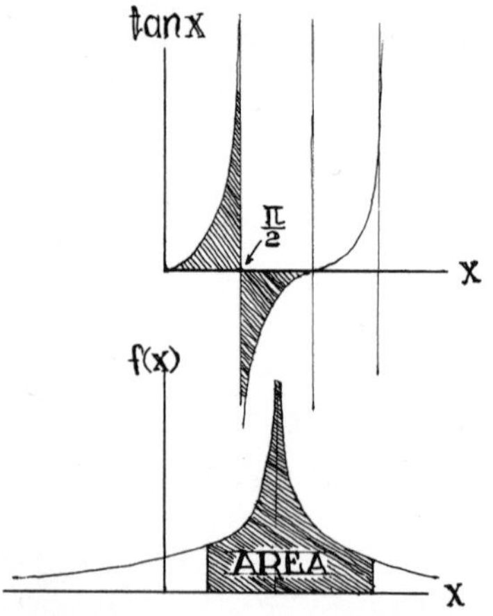

One final means of characterizing a singularity, is to measure the area under the curve in an interval containing the singularity.

There are but two types of curves relative to this criterion; those having a finite area under the singularity and those having an infinite area under the singularity.

G. Areas

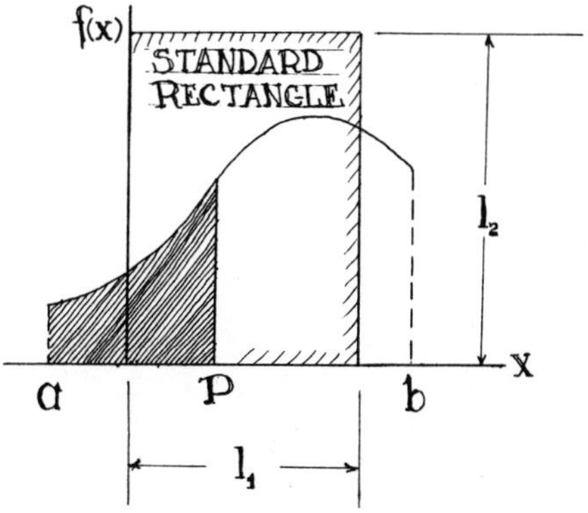

Regard the curve representing the function f(x) defined in the interval $a \leq x \leq b$. The area under the curve between x = a and x = P is shown as the shaded region.

This area for practical purposes at this point can be obtained by various approximate techniques.

Approximation Methods

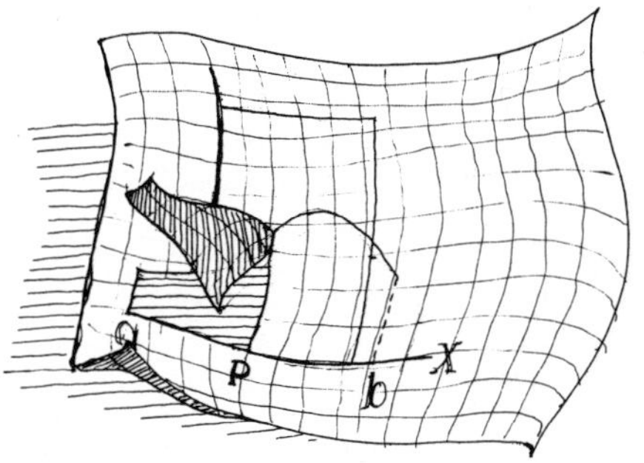

First plot the curve on paper of constant density and thickness and then cut out a standard rectangular area and also cut out the area in question. By comparing the ratio of the weights of the **two cutouts** we can obtain the area under the function from x = a to x = P. If the sides of the standard rectangle are l_1 and l_2, then the area is $l_1\ l_2$.

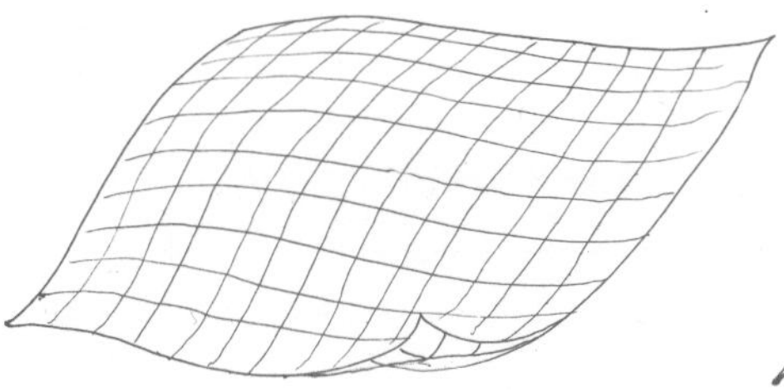

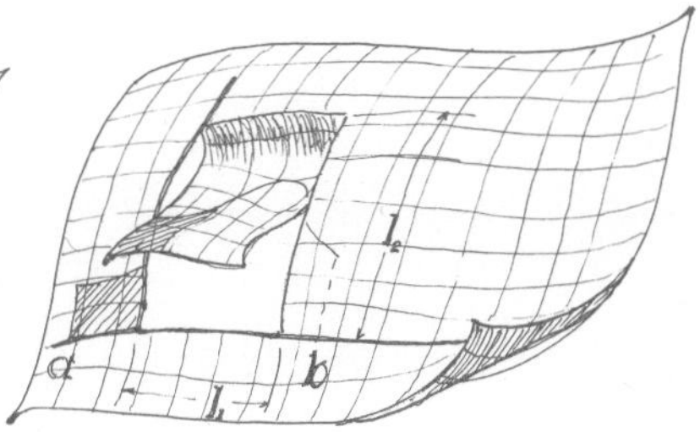

Therefore,

$$\frac{\text{Weight of area under f(x) from x = a to x = P}}{\text{Weight of rectangular area } (l_1 \cdot l_2)} = \frac{\text{area under f(x)}}{(l_1 \cdot l_2)}.$$

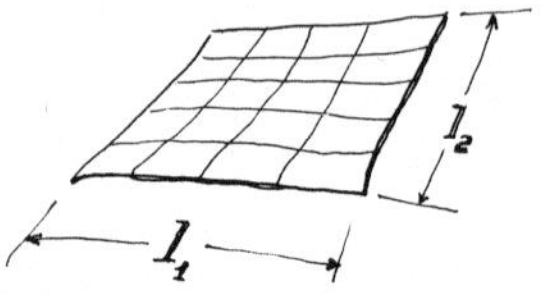

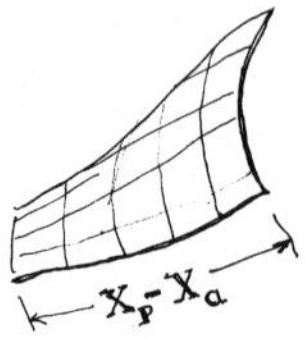

If f(x) is plotted upon cross hatched paper we can count squares in the region in question.

Rectangular Partitioning

This approximation method for finding the area under a curve is extremely useful. Thus it is worth our while to examine this method in detail.

Our problem is to obtain an approximation to the area under the curve f(x) between the point a and a point x = P.

To do this let us divide the interval a → P into n **equal segments** of width ε; thus

$$\epsilon = \frac{P - a}{n}$$

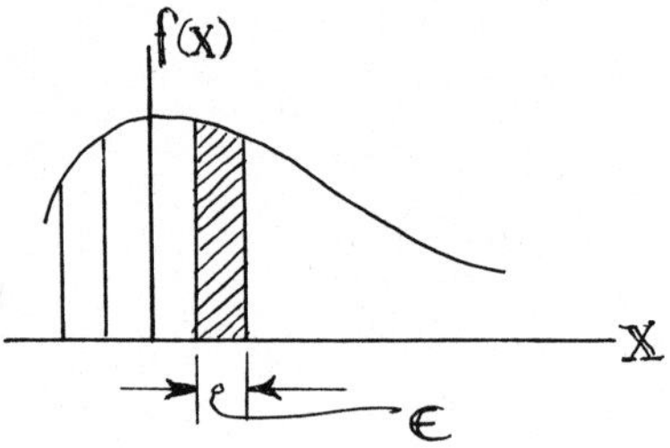

We set ε **small enough** that the slope of f(x) does not change by great amount in the interval ε.

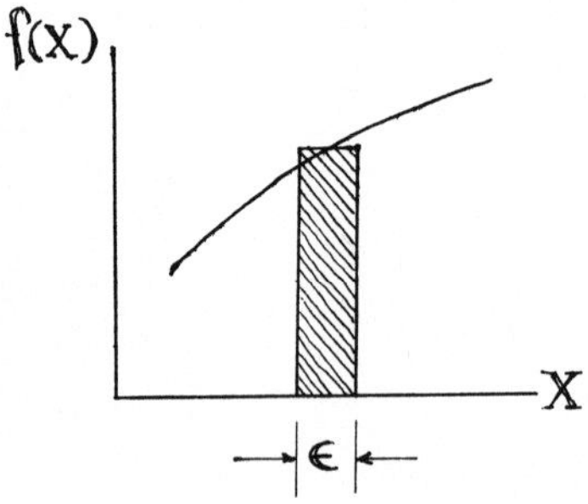

slope relatively constant in the interval ε.

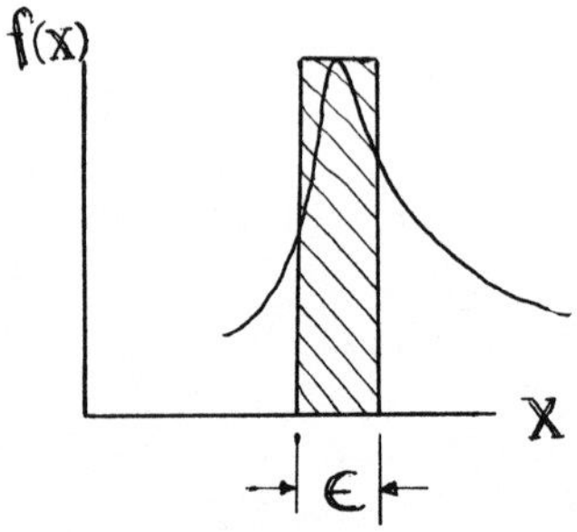

ε is too large in this case. The slope changes decidedly in the interval.

The method of **rectangular partitioning** is performed by passing a horizontal line through f(x) at the **mid point** of *each* interval ϵ. The horizontal lines when joined to the vertical lines at the extremities of the interval ϵ form a **rectangle.**

This procedure is illustrated below.

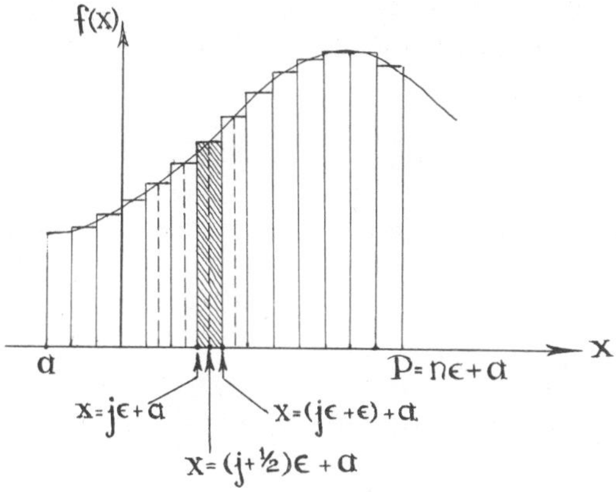

Regard the jth interval.

The area of the jth interval $= \epsilon \cdot f(a + [j - 1/2]\,\epsilon)$.

The total area between a and P can be obtained by summing the n rectangles.

$$\text{Total Area, } a \rightarrow P = A_{a\rightarrow p} \cong \sum_{j=1}^{j=n} \epsilon \cdot f(a + [j - 1/2]\,\epsilon)$$

To illustrate this let us take a case in which a = 0 and P = 10. Assume 10 intervals.

$$n = 10 \text{ and } \epsilon = 1$$

Then we make a table.

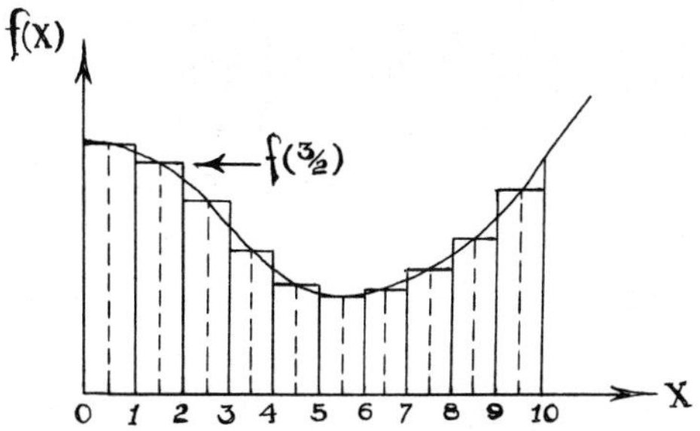

j	j − ½	f(a + [j − ½] ε)
1	½	f(½)
2	3/2	f(3/2)
3	5/2	f(5/2)
4	7/2	f(7/2)
5	9/2	f(9/2)
6	11/2	f(11/2)
7	13/2	f(13/2)
8	15/2	f(15/2)
9	17/2	f(17/2)
10	19/2	f(19/2)

If we now add all of the $f(a + (j - 1/2)\epsilon)'$ s and **multiply by ϵ** we have our **approximation.**

$$A_{0-10} \simeq \epsilon \cdot \{f(1/2) + f(3/2) + \cdots\cdots + f(19/2)\}$$

A simple **numerical example** will further clarify our method. In the case of the parabola $f(x) = 2x^2$ find the area from x = o to x = 1 using **two intervals,** i.e.

$$n = 2; \epsilon = 1/2$$

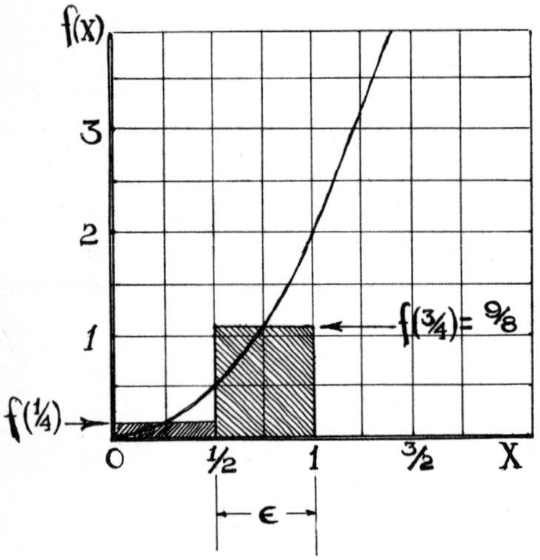

Our table becomes

j	j − ½	f(a + [j − ½] ε)
1	½	f(¼) = ⅛
2	3/2	f(¾) = 9/8

and

$$A_{0\to 1} \simeq \frac{1}{2}\left\{\frac{1}{8} + \frac{9}{8}\right\} = 5/8$$

The true value of this area by **integration** is

$$A_{0\to 1} = \frac{2}{3}$$

The fraction error in our calculation then is

$$\frac{2/3 - 5/8}{2/3} = \frac{1}{16}$$

Trapezoidal Partitioning

It should be a reasonable assumption that once the interval $a \to P$ is split up into n **sub-intervals** of width ϵ, various connecting curves can be used to form the top of the area defined by ϵ.

The same considerations concerning changes in slope over the interval ϵ still apply.

Another useful partition method is that of forming trapezoids. The points at which the extremities of the interval ϵ cross f(x) can be connected by **straight line segments** to form areas of the shape shown below. If the curvature has predominantly one sign in the interval this method either overestimates or underestimates the area. Whether the evaluation is high or low depends on the predominant sign of the curvature.

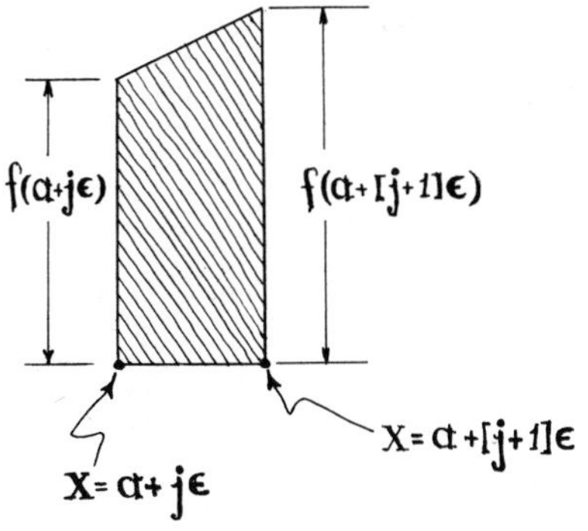

This shape is a trapezoid of area

$$\left\{\frac{f(a + j\epsilon) + f(a + [\,j + 1\,]\,\epsilon)}{2}\right\} \cdot \epsilon$$

A typical **trapezoidal partitioning** is shown below.

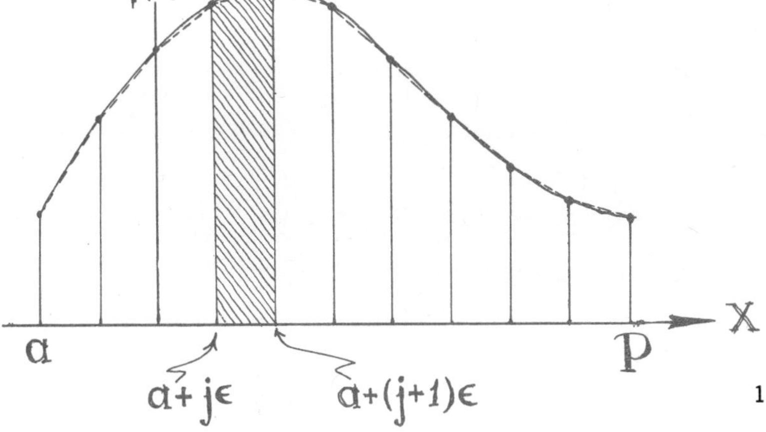

In this case

$$A_{a \to P} \simeq \epsilon \cdot \left\{ \frac{f(a) + f(a + \epsilon)}{2} + \frac{f(a + \epsilon) + f(a + 2\epsilon)}{2} + \dots \right.$$

$$\left. + \dots \frac{f(a + (n - 1)\epsilon) + f(a + n\epsilon)}{2} \right\}$$

$$= \epsilon \cdot \left\{ \sum_{j=0}^{j=n} f(a + j\epsilon) - \frac{1}{2} f(a) - \frac{1}{2} f(a + n\epsilon) \right\}$$

To illustrate this method let us approximate the area under the parabola used before in the interval, 0 to 1.

$$f(x) = 2x^2$$

Then if $n = 2, \epsilon = 1/2$.

X	j	$f(a + j\epsilon)$
0	0	$f(0) = 0$
½	1	$f(½) = ½$
1	2	$f(1) = 2$

$$A_{0 \to 1} \simeq \frac{1}{2} \left\{ \frac{f(o) + f(1/2)}{2} + \frac{f(1/2) + f(1)}{2} \right\}$$

$$= \frac{1}{4} \left\{ \frac{1}{2} + 5/2 \right\} = 3/4$$

$$\text{The fractional error} = \frac{2/3 - 3/4}{2/3} = -\frac{1}{8}$$

We notice that the method of rectangular partitioning is better than that of trapezoidal partitioning for *this particular function.* This large error for the latter method is caused by the overall positive curvature of the function.

The choice of method is a matter of judgment depending on the special problem to be solved.

The Area under the Slope Curve

In this section we will prove a very important theorem.

The **area** under the slope curve of a function f(x) obtained by cord approximation in the interval from x = a to a point x is **equal** to the **value** of the function at x, f(x) minus a constant. The constant is f(a).

The approach is quite straightforward.

1. We first will obtain the approximate slope curve d(x) or tan α_x by the cord approximation in the interval ϵ.
2. Second we will find the area under the approximate slope curve from a to x by **rectangular partitioning.**

As before let us first construct our function f(x) in the interval $a \leq x \leq b$, splitting the interval into n **sub-intervals** of width ϵ.

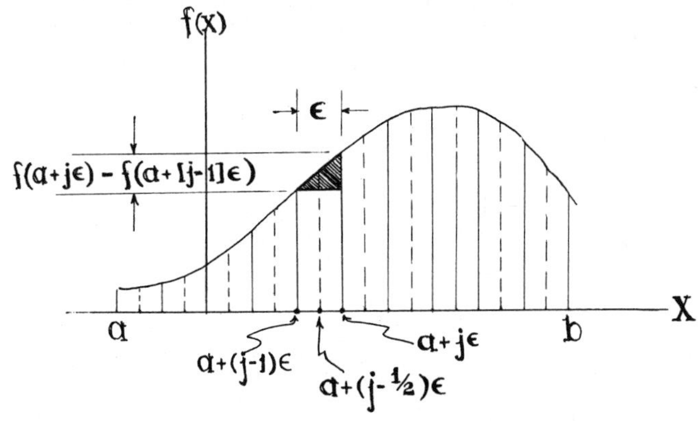

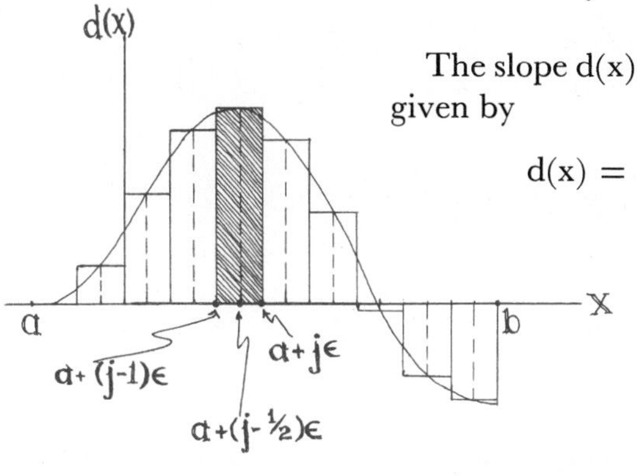

The slope d(x) at every **half interval point** is given by

$$d(x) = d(a + (j - \frac{1}{2})\epsilon) = \frac{f(a + j\epsilon) - f(a + (j - 1)\epsilon)}{\epsilon}$$

Fortunately we have arranged our intervals in a convenient manner since

$$[\text{Area Slope Curve}]_{a \to x} \simeq \sum_{j=1}^{j=n} \epsilon \cdot d\,(a + (j - 1/2)\,\epsilon)$$

but

d (a + (j − 1/2) ϵ) from our approximation is

approximately

$$\frac{f\,(a + j\epsilon) - f(a + (j - 1)\,\epsilon)}{\epsilon}$$

Then

$$[\text{Area Slope Curve}]_{a \to x} \simeq \sum_{j=1}^{j=n} \epsilon \left\{ \frac{f(a + j\epsilon) - f(a + (j - 1)\,\epsilon)}{\epsilon} \right\}$$

$$\simeq \sum_{j=1}^{j=n} \{f(a + j\epsilon) - f(a + (j - 1)\epsilon)\}$$

$$\begin{aligned} &\simeq \{f(a + \epsilon) - f(a) + f(a + 2\epsilon) - f(a + \epsilon) + \ldots \\ &\quad + f(a + (n - 1)\epsilon) - f(a + (n - 2)\epsilon) \\ &\quad + f(a + n\epsilon) - f(a + (n - 1)\epsilon)\} \\ &\simeq f\,(a + n\epsilon) - f\,(a) \end{aligned}$$

Since $x = (a + n\epsilon)$

$$[\text{Area Slope Curve}]_{a \to x} \simeq f\,(x) - f(a) = f(x) - \text{a constant}$$

We are left with the final extension in the integral calculus to show that this is an exact equality.

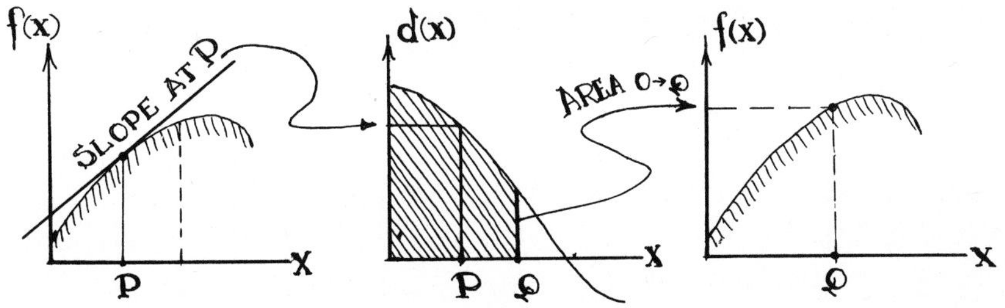

It is instructive to illustrate this with a simple numerical example. If we make use of the parabola

$$f(x) = \frac{1}{10} x^2$$

in the interval from

$$0 \leq x \leq 10$$

take *Five* **intervals** from 0 to 10

$$\epsilon = \frac{10 - 0}{5} = 2$$

knowing the values of tan α_x or d(x) as function of x we can compute the area under the slope curve from x = 0 to x = 6.

$$[\text{Area}]_{0 \to 6} = \epsilon \cdot \sum_{j=1}^{j=3} d(0 + (j - 1/2)\epsilon)$$

$$= 2 \{d(1/2) + d(3/2) + d(5/2)\}$$

$$= 2 \left\{ \frac{2}{10} + \frac{6}{10} + \frac{10}{10} \right\} = 2 \left\{ \frac{18}{10} \right\} = \frac{36}{10} = \frac{18}{5}$$

Now $-f(a = o) + f(x = 6) = -0 + \frac{36}{10} = \frac{18}{5}$

This result demonstrates our theorem.

Area Computations in Terms of Matrices

Earlier the slope of a curve f(x) was represented as a linear transformation. Previous to this section the area computation was shown to be the inverse of the slope calculation. This **inverse relationship** becomes even more apparent when the area calculations are represented in terms of a linear transformation.

To a reasonable approximation the area between the x axis and f(x) in the interval 0 to $n\epsilon$ can be represented as

$$\Lambda(n\epsilon) = \text{area from 0 to } n\epsilon \simeq \sum_{j=1}^{j=n} \epsilon \cdot f(j\epsilon)$$

Write f(x) as a column vector having its jth element given as

$$f_j = f(j\epsilon)$$

with

$$\mathbf{f} = \begin{bmatrix} f(\epsilon) \\ f(2\epsilon) \\ \vdots \\ f(N\epsilon) \end{bmatrix} = \begin{bmatrix} f_1 \\ f_2 \\ \vdots \\ f_N \end{bmatrix}$$

Also let Λ (x) (the area between 0 and x) be a vector whose elements are

$$\Lambda_n = \Lambda(n\epsilon)$$

where

$$\boldsymbol{\Lambda} = \begin{bmatrix} \Lambda(\epsilon) \\ \Lambda(2\epsilon) \\ \vdots \\ \Lambda(N\epsilon) \end{bmatrix} = \begin{bmatrix} \Lambda_1 \\ \Lambda_2 \\ \vdots \\ \Lambda_N \end{bmatrix}$$

Then

$$\Lambda_n = \sum_{j=1}^{n} A_{nj}\, f_j$$

and

$$A_{nj} = \epsilon \qquad j \leq n$$
$$A_{nj} = O \qquad n > j$$

With this approximation the area matrix $\mathbb{A}$ assumes a familiar form,

$$\mathbb{A} = \epsilon \begin{pmatrix} 1 & O & O & O & O & O \\ 1 & 1 & O & O & O & O \\ 1 & 1 & 1 & O & O & O \\ 1 & 1 & 1 & 1 & O & O \\ 1 & 1 & 1 & 1 & 1 & O \\ 1 & 1 & 1 & 1 & 1 & 1 \end{pmatrix}$$

$\mathbb{A}$ is the **triangular matrix** and as such is the **inverse** of the slope matrix $\mathbb{D}$;

$$\mathbb{A} = \mathbb{D}^{-1}$$

Note carefully that we are working with an approximation method. Thus the results obtained here can only be considered as approximate relationships. Later we will observe that the operation of integration (area computation) is the inverse of differentiation (slope computation) in ONE definition of the integral.

It is interesting to observe that both operations involve half interval shifts, however the shift of $\mathbb{A}$ is compensated for in the reverse shift of $\mathbb{D}$.

Using the techniques available at this point we can solve simple problems. As an example solve the following problem.

The slope of the function f(x) in the interval $x = 0$ to $x = 8$ is given by $g(x) = x$. Find f(x).

Symbolically this problem is written

$$\mathbb{D} \cdot \mathbf{f} = \mathbf{g}$$

or

$$\mathbb{D} \cdot \begin{bmatrix} f_1 \\ f_2 \\ \vdots \\ f_N \end{bmatrix} = \begin{bmatrix} g_1 \\ g_2 \\ \vdots \\ g_N \end{bmatrix}$$

Take N as 8 then $\epsilon = 1$ and $g(n\epsilon) = n\epsilon$

$$\mathbf{g} = \begin{bmatrix} \epsilon \\ 2\epsilon \\ 3\epsilon \\ \vdots \\ 8\epsilon \end{bmatrix} = \begin{bmatrix} 1 \\ 2 \\ 3 \\ \vdots \\ 8 \end{bmatrix}$$

The problem is solved quite readily for **f** by clearing with $\mathbb{A}$ or $\mathbb{D}^{-1}$

$$\mathbb{A} \cdot \mathbb{D} \cdot \mathbf{f} = \mathbb{I} \cdot \mathbf{f} = \mathbf{f} = \mathbb{A} \cdot \mathbf{g}$$

and

$$\begin{bmatrix} f_1 \\ f_2 \\ f_3 \\ \vdots \\ \vdots \\ f_8 \end{bmatrix} = \begin{pmatrix} 1 & O & O & O & O & O & O & O \\ 1 & 1 & O & O & O & O & & \\ 1 & 1 & 1 & O & O & & & \\ 1 & 1 & 1 & 1 & \ddots & & & \\ 1 & 1 & & & & & & \\ \vdots & & & & & & \ddots & \\ \cdots & \cdots & \cdots & \cdots & \cdots & \cdots & \cdots & \cdots \end{pmatrix} \begin{bmatrix} 1 \\ 2 \\ 3 \\ \vdots \\ \vdots \\ 8 \end{bmatrix}$$

After carrying out the matrix multiplication

$$\mathbf{f} = \begin{bmatrix} 1 \\ 3 \\ 6 \\ 10 \\ \vdots \\ 36 \end{bmatrix}$$

Again keep in mind that in working out problems using these techniques constant care must be exercised with respect to values obtained at the end points.

CHAPTER 5

Differential and Integral Calculus

A. Introduction

IN THE PREVIOUS CHAPTER the computation of the approximate *area* under the curve representing the slope at every point of a function f(x) was shown to be related approximately to the function from which the slope was taken. This class of computations is the major interest of the differential and integral calculus.

The derivative of a function is a measure of the slope of that function at a point in the limit as the interval considered becomes arbitrarily small.

There are a number of ways of defining the integral. In the discussion to follow the integral will be considered as the anti-derivative or the inverse operation to that of differentiation.

The integral of a function f(x) is a measure of the AREA between the function and the x axis. Areas above the axis are positive and areas below the axis are negative.

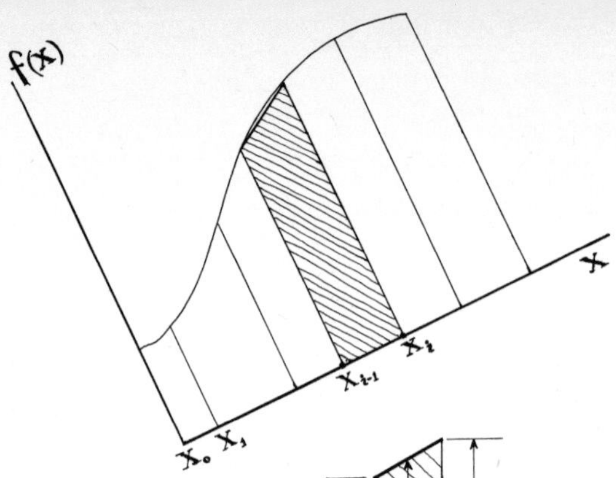

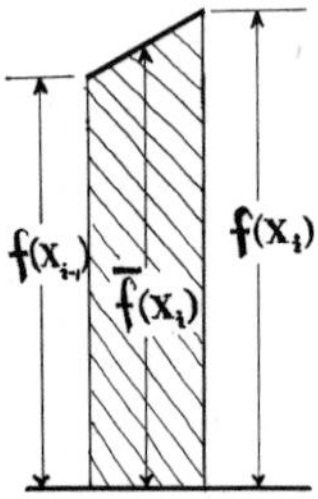

The diagram shown here will serve to introduce the approach.

The basic relationship between the **derivative** and the **integral** will be demonstrated with this simple diagram of a trapezoidal element.

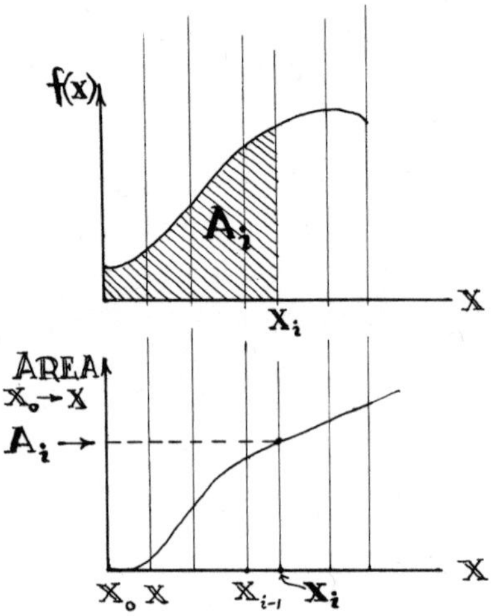

Instruction! Plot the area under the curve f(x) from 0 to x as a function of x.

Let A_{i-1} be the area from

O to x_{i-1}

Let A_i be the area from

O to x_i

Definition!

The symbol ΔA_i will represent the difference between the area A_i and the area A_{i-1}.

$$\boxed{\Delta A_i = A_i - A_{i-1}}$$

Since A_i is denoted by the index at the right of the interval, the **total area** to a point x_j is given by

$$A_j = \sum_{m=1}^{j} \Delta A_m = \Delta A_1 + \Delta A_2 + \ldots \Delta A_j$$

$$= (A_0 - A_1) + (A_1 - A_2) + \ldots (A_{j-1} - A_j)$$

As a special case take the function

$$f(x) = 1$$

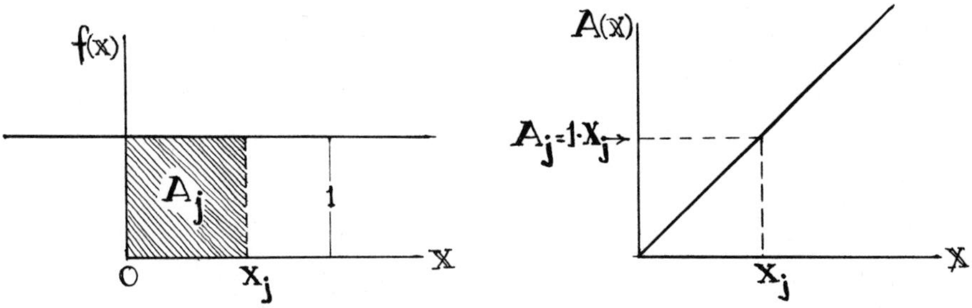

The area of the rectangle from x = o to x is

$$A(x) = 1 \cdot x$$

We can therefore compute and plot A(x) as a function of x.

This is a particularly simple example in which the slope of A(x) is constant and equal to one. It is apparent here that the slope of A(x) equals f(x).

The symbol Δ is used consistently to signify differences such as

$$x_{i-1} - x_i = \Delta x_i$$

The term $\Delta A_i = A_i - A_{i-1}$ from the area curve is given approximately from the trapezoidal element of the f(x) diagram.

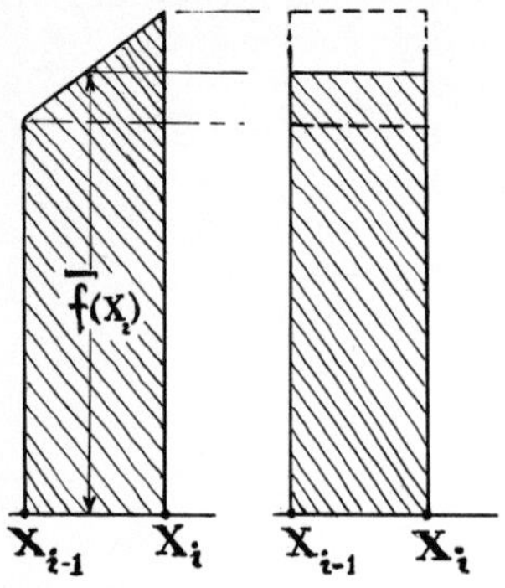

Define! $\bar{f}(x_i) = \frac{1}{2}\{f(x_i) + f(x_{i-1})\}$

= the mean value in the interval Δx_i

If the interval $\Delta x_i = x_i - x_{i-1}$ is sufficiently **small** the arc of f(x) in this interval can be considered to be approximately a straight line.

Then

$$\Delta A_i = A_i - A_{i-1} = \frac{1}{2}\{f(x_i)\,\Delta x_i + f(x_{i-1})\,\Delta x_i\}$$

$$= \frac{1}{2}\{f(x_i) + f(x_{i-1})\}\,\Delta x_i$$

$$= \bar{f}(x_i)\,\Delta x_i$$

In other words we have calculated the mean rectangular area.

By dividing both sides of this equation by Δx_i we can relate the average value of the function f(x) in the interval Δx_i to the slope of the area curve.

$$\frac{\Delta A_i}{\Delta x_i} = \bar{f}_i$$

where $\bar{f}_i = \bar{f}(x_i)$

The net result of this is that the area (as we would expect) is *approximately* the sum of the areas of the trapezoids $\bar{f}(x_i)\,\Delta x_i$.

$$A_j = \sum_{i=1}^{i=j} \Delta A_i = \sum_{i=1}^{i=j} \bar{f}(x_i)\,\Delta x_i$$

The integral of f(x) from x_0 to x_j is written as

$$A_j = \int_0^{x_j} f(x)dx = \underset{\Delta x_i \to 0}{\text{Limit}} \sum_{i=0}^{i=j} f(x_i)\,\Delta x_i$$

where the number of intervals between o and x_j becomes appropriately large as the width of each Δx_i approaches zero.

In other words

$$\underset{\Delta x_i \to o}{\text{Limit}} \sum_{i=1}^{i=j} \Delta x_i = x_j - x_0 = \text{length along x between zero and } x_j$$

The number of intervals must be such that the sum of Δx_i's gives the length $x_j - x_0$.

We have introduced the term **LIMIT.**

There are subtle problems involved in the consideration of the process of taking a **limit;** therefore we will leave the more extensive discussion of this subject to a separate section.

We defined the **integral** as the **limit** of a **sum** of ΔA_i's.

The derivative of A(x) will be defined as the limit of the ratio of ΔA_i to Δx_i as Δx_i approaches zero.

The derivative with respect to x is denoted by the symbol

$$\frac{d}{dx}$$

$$\text{Derivative of } A(x) = \frac{dA(x)}{dx} = \underset{\Delta x_i \to o}{\text{Limit}} \frac{\Delta A_i}{\Delta x_i}$$

Finally we note the anti-derivative characteristic of the integral.

If

$$\bar{f}(x_i) \simeq \frac{\Delta A_i}{\Delta x_i}$$

then

$$\underset{\Delta x_i \to o}{\text{Limit}}\ \bar{f}(x_i) = f(x_i) = \frac{dA_i}{dx_i}$$

The reader can rightly claim that more has been implied here than has been discussed.

What is the point?

As an introduction we have demonstrated that

$$\bar{f}(x_i) \simeq \frac{\Delta A_i}{\Delta x_i}$$

or that the *slope* of the *area curve* A(x) is approximately *equal* to the *average value* of the *function* f(x) in the interval of interest.

Furthermore we hope to demonstrate the equality

$$f(x) = \frac{dA(x)}{dx}$$

by noting that the **area** given by the sum

$$A_j = \underset{\Delta x_j \to 0}{\text{Limit}} \sum_{i=1}^{j} \Delta A_i = \int_0^{A_j} dA$$

$$= \underset{\Delta x_i \to 0}{\text{Lim}} \sum_{i=1}^{j} \frac{\Delta A_i}{\Delta x_i} \Delta x_i = \int_0^{A_j} \frac{dA}{dx} dx$$

$$= \underset{\Delta x_i \to 0}{\text{Lim}} \sum_{i=1}^{j} \bar{f}(x_i) \Delta x_i = \int_{X_0}^{X_j} f(x)\, dx$$

Thus

$$f(x) = \frac{dA(x)}{dx}$$

With this general survey of what is to follow we can turn to a detailed consideration of the **limit** followed by a development of the **derivative** and the **integral** (or **anti-derivative**).

B. *The Concept of a Limit*

Limits of Various Example Functions

The notion of a "limiting value" introduced in the preceding section is of critical importance in mathematical analysis. We shall examine the notion briefly without specific reference to areas or tangent lines.

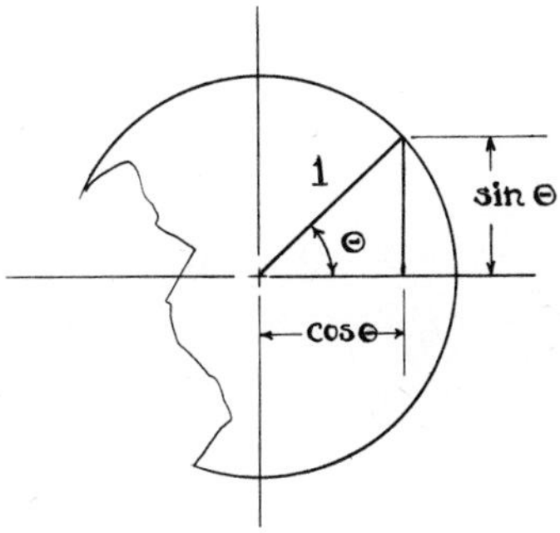

1. The Sine and Cosine.

In the diagram to the right we represent the sine and cosine functions geometrically.

If $\theta \to O$, then the vertical segment representing $\sin \theta$ also approaches zero.

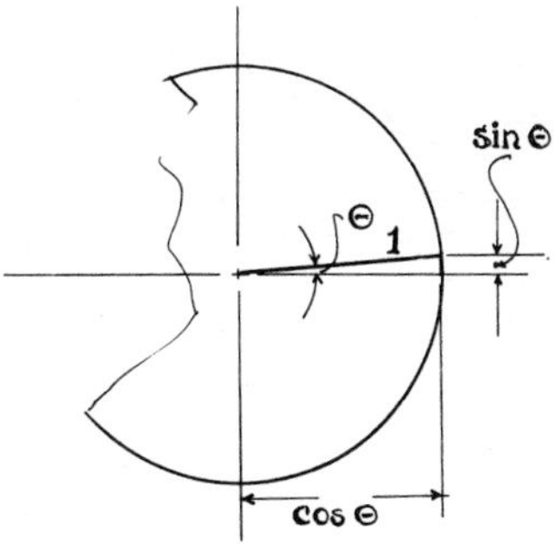

$$\underset{\theta \to O}{\text{Limit}} \sin \theta = O$$

and conversely. In addition

$$\text{Cos}\,\theta = \sqrt{1 - \text{Sin}^2\,\theta}$$

hence as $\theta \to O$

$$\text{Cos}\,\theta \to 1$$

These statements although elementary are useful as illustrations since they are geometrically suggested.

2. Consider the Rational Function.

$$\frac{(x^2 + 3x + 1)}{(2x^2 - x)}$$

As $x \to \infty$ (or as $\frac{1}{x} \to 0$) the expression above tends to the value 1/2. This value is evident if we rewrite the rational function above as

$$\frac{(1 + 3/x + 1/x^2)}{(2 - 1/x)}$$

The terms $3/x$, $1/x^2$, and $1/x$ all approach zero as x is made arbitrarily large.

3. The limit of the **sequence.**

$$S_n = 1 + x + x^2 + \ldots x^n$$

approaches the value $\frac{1}{1 - x}$ (in the closed interval ($|x| < 1$) as n is made arbitrarily large. This is usually written "as $n \to \infty$".

This result is demonstrated by multiplying S_n by x and subtracting S_n

$$xS_n - S_n = (x + x^2 + \cdots x^{n+1}) - (1 + x + \cdots x^n)$$
$$S_n(x - 1) = S_{n+1} - 1 - S_n = -(1 - x^{n+1})$$

Therefore

$$S_n = \frac{(1 - x^{n+1})}{(1 - x)} = \frac{1}{(1 - x)} - \frac{x^{n+1}}{(1 - x)}$$

If $|x| < 1$ then $|x|^{n+1} \to O$ as $n \to \infty$

and

$$\underset{n \to \infty}{\text{Limit}}\, S_n = \underset{n \to \infty}{\text{Limit}} \frac{(1 - x^{n+1})}{(1 - x)} = \frac{1}{1 - x}$$

4. The function $\frac{\text{Sin } x}{x} \to 1$ as $x \to O$. This statement can be demonstrated geometrically or analytically.

If we represent sin x as a series expansion

$$\text{Sin } x = \sum_{n=0}^{\infty} \frac{(-1)^n x^{(2n+1)}}{(2n+1)!}$$

$$= x - \frac{x^3}{3!} + \frac{x^5}{5!} + \text{------}$$

then

$$\frac{\sin x}{x} = 1 - \frac{x^2}{3!} + \frac{x^4}{5!} \quad \text{---------}$$

and

$$\underset{x \to 0}{\text{Limit}} \frac{\sin x}{x} = 1$$

The Formal Definition of a Limit

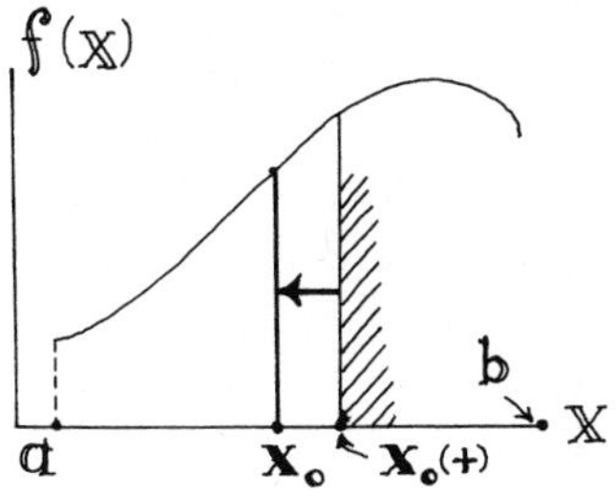

With these examples in view we proceed to a more formal definition of a Limit.

Let f(x) be a function (possibly many valued) defined for all x in some interval $x_0 < x < b$. Note that the interval can be open; in other words we allow the possibility that f(x) may not be defined at x_0 (or b).

We write

$$\lim_{x \to x_0(+)} f(x) = L$$

in this expression $x_0(+)$ is a value of x to the right of x_0 and arbitrarily close to x_0.

$$L - f(x) \to 0 \text{ as } x \to x_0(+)$$

i.e., if $L - f(x)$ can be made arbitrarily small in absolute value by **confining** x to a small enough interval on the right of x_0, say $x_0 < x < (x_0 + \Delta x)$, then we call L,

The limit from the right of f(x) as $x \to x_0$ from the right.

In a similar manner, if f(x) is also defined to the left of x_0, say in the interval $a < x < x_0$, we can define the **left-hand limit:**

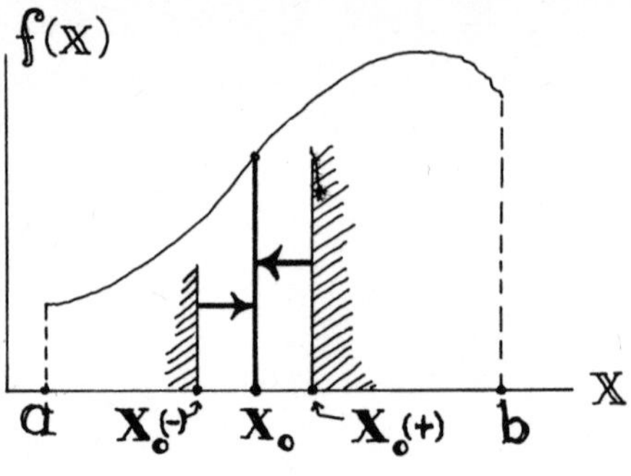

$$\lim_{x \to x_0(-)} f(x) = L'$$

This equation means that $L' - f(x)$ can be made arbitrarily small in magnitude by confining x to vary in a sufficiently small interval

$$x_0 - \Delta x < x < x_0$$

to the **left** of x_0.

If $L = L'$ we may write simply

$$\lim_{x \to x_0} f(x) = L \text{ or } L'$$

As an example of a function which has neither right nor left-handed limits we examine

$$f(x) = \sin\frac{1}{x} \qquad \text{as } x \to O$$

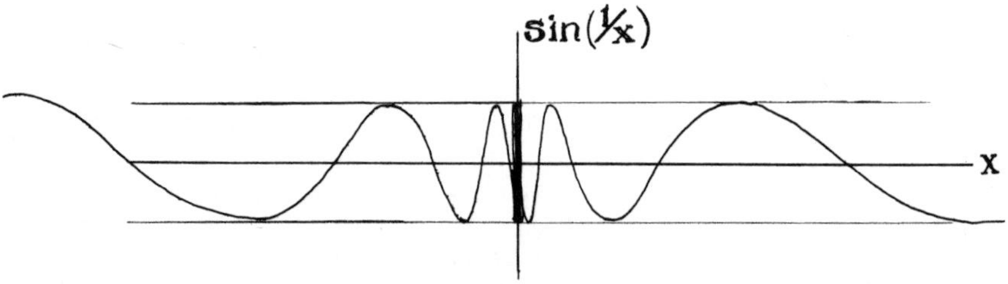

This particular function oscillates more and more rapidly as x approaches zero from either side, and the function is undefined at $x = O$.

In our definition of $\lim_{x \to x_0} f(x)$ we have assumed that x_0 is finite. We extend the definition of the **symbol**

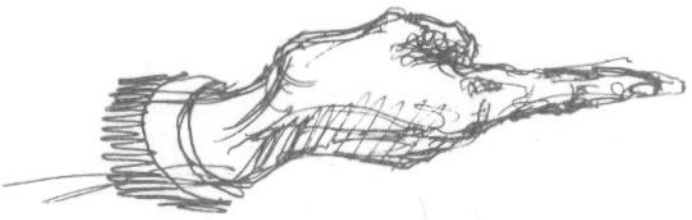

LIMIT

by $\lim_{x \to +\infty} f(x)$ to mean $\lim_{\left(\frac{1}{x}\right) \to 0(+)} f(x)$

and

$\lim_{x \to -\infty} f(x)$ to mean $\lim_{\left(\frac{1}{x}\right) \to 0(-)} f(x)$

In our previous example of the sequence

$$S_n = 1 + x + x^2 + x^n$$

the quantity S_n is viewed as a function of n where n takes on only *positive integral* values. For a function f(n) in which the variable can take on only integral values we make the definition of the limit analogous to those above: vis;

$$\lim_{\left(\frac{1}{n}\right) \to O(+)} f(n) = \lim_{n \to \infty} f(n) = L$$

if $L - f(n)$ can be made arbitrarily small by confining $\left(\frac{1}{n}\right)$ to an arbitrarily small interval

$$O < \left(\frac{1}{n}\right) < \delta, \text{ (where } \delta \text{ can be made arbitrarily small),}$$

or what comes to the same thing by confining n to suitably large values

$$n_0 < n < \infty$$

We are able to construct a similar definition for

$$\lim_{n \to -\infty} f(n)$$

Continuity

We can now give a more precise definition of continuity. Let f(x) be defined and single-valued at every point of an interval $a < x < b$ (or $a \leq x \leq b$, etc.). Then f(x) is *continuous* in the interval if at each x in the interval we have

$$\lim_{x' \to x} f(x') = f(x)$$

(At the end-points we use the one-sided limits: $\lim_{x' \to a(+)} f(x') = f(a)$, etc.)

It is easily seen that the examples above conform to the definitions we have just made.

Simple formal properties.

$$\lim (f \cdot g) = (\lim f)(\lim g)$$
$$\lim (f \pm g) = (\lim f) \pm (\lim g)$$
$$\lim (f/g) = \frac{\lim f}{\lim g}$$

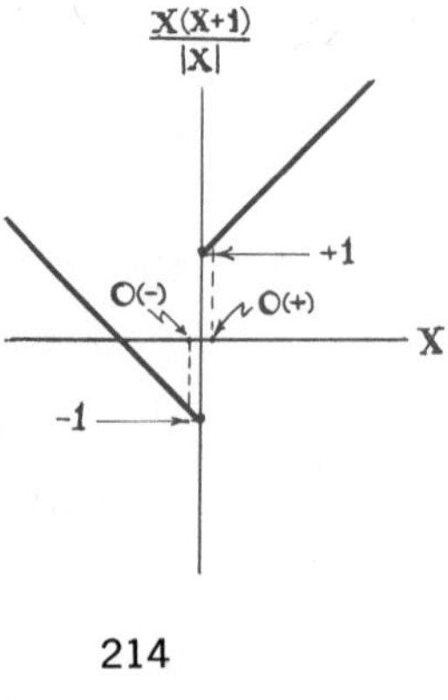

Why all of this worry?
Some functions are discontinuous.

As an example of a discontinuous function consider $f(x) = \dfrac{x(x+1)}{|x|}$.
For $x > O$ we have $|x| = x$, hence $f(x) = x + 1$, and so $f(x) \to 1$ as $x \to O(+)$. For $x < O$ we have $f(x) = -(x + 1)$, and so $f(x) \to -1$ as $x \to O(-)$. The function is not defined for $x = O$. The graph is shown at the left.

Fundamental Properties of Limits

Suppose f(x) and g(x) are defined for $x_0 < x < b$, and suppose that

$$f(x) \to L \text{ and } g(x) \to L' \text{ as } x \to x_0(+)$$

or, in the earlier notation,

$$\lim_{x \to x_0(+)} f(x) = L \quad \text{and} \quad \lim_{x \to x_0(+)} g(x) = L'$$

We claim that

$$\begin{aligned} &f(x) \pm g(x) \to L \pm L' \text{ as } x \to x_0(+) \\ &f(x) \cdot g(x) \to L \cdot L' \quad \text{as } x \to x_0(+) \\ &f(x)/g(x) \to L/L' \quad \text{as } x \to x_0(+), (\text{if } L' \neq O) \end{aligned}$$

The verification of the first of these we leave as an exercise. For the second and third put

$$\epsilon(x) = f(x) - L \qquad\qquad \eta(x) = g(x) - L'$$

so that, by definition, $|\epsilon(x)|$ and $|\eta(x)|$ can be made arbitrarily small by confining x to a sufficiently small interval to the right of x_0, say $x_0 < x < (x_0 + \Delta x)$. We must show that $|f(x)\, g(x) - LL'|$ and $|f(x)/g(x) - L/L'|$ can be made arbitrarily small by confining x to a sufficiently small interval to the right of x_0. Now

$$\begin{aligned} f(x)\, g(x) &= (\epsilon(x) + L) \cdot (\eta(x) + L') \\ &= \epsilon \cdot \eta + \epsilon \cdot L' + \eta \cdot L + LL' \end{aligned}$$

and so

$$\begin{aligned} |f(x) \cdot g(x) - LL'| &= |\epsilon \cdot \eta + \epsilon \cdot L' + \eta \cdot L| \\ &\leq |\epsilon||\eta| + |\epsilon||L'| + |\eta||L| \end{aligned}$$

Clearly the right-hand side, hence also the left, can be made as small as desired by keeping $x_0 < x < (x_0 + \Delta x)$ with small enough Δx, since that is true of ϵ and η.

Similarly,

$$\frac{f}{g} - \frac{L}{L'} = \frac{\epsilon + L}{\eta + L'} - \frac{L}{L'} = \frac{\epsilon L' + LL' - L\eta - LL'}{\eta L' + L^{2'}}$$

$$= \frac{\epsilon L' - \eta L}{\eta L' + (L')^2}$$

Therefore

$$|f/g - L/L'| = \frac{|\epsilon L' - \eta L|}{|\eta L' + (L')^2|}$$

Clearly the numerator on the right can be made as small as desired by keeping $x_0 < x < (x_0 + \Delta x)$, with sufficiently small Δx. Assuming $L' \neq O$ then $L^{2'} \neq O$, and the denominator on the right cannot get arbitrarily small when x is near enough to x_0. Q. E. D.

These properties can equally well be proved for left-hand limits and for limits in which the independent variable x can only take on whole number values. We leave the verification of these modifications to the reader. Briefly stated, the above rules simply say that the *limit of a sum (or difference, product, quotient)* of *two functions is equal to the sum (or difference, product, quotient of the limits)*. The rules can easily be extended to any sum, product, etc., of a *finite* number of functions.

An application. If f(x) is continuous in an interval $a < x < b$, then by definition, $\lim_{\Delta x \to 0} f(x + \Delta x) = f(x)$ for every x in the interval. If also g(x) is continuous in $a < x < b$, then, applying the above limit rules, so are $(f \pm g)$, $(f \cdot g)$, and also (f/g) *except at points where* $g = O$.

Now $f(x) = x$ is clearly continuous. Therefore $f(x) \cdot f(x) = x^2$ is continuous for all values of x. Applying the same rule repeatedly we see that $f(x) = x^n$ is continuous for all x and for any $n = 1, 2, 3, 4$, etc. Also any constant function is obviously continuous. We conclude from our rules that any polynomial function

$$f(x) = a_0 + a_1 x + \ldots . + a_n x^n$$

is continuous, and finally we conclude that any rational function

$$f(x) = \frac{a_0 + a_1 x + \ldots + a_n x^n}{b_0 + b_1 x + \ \ldots + b_m x^m}$$

is continuous except where the denominator vanishes.

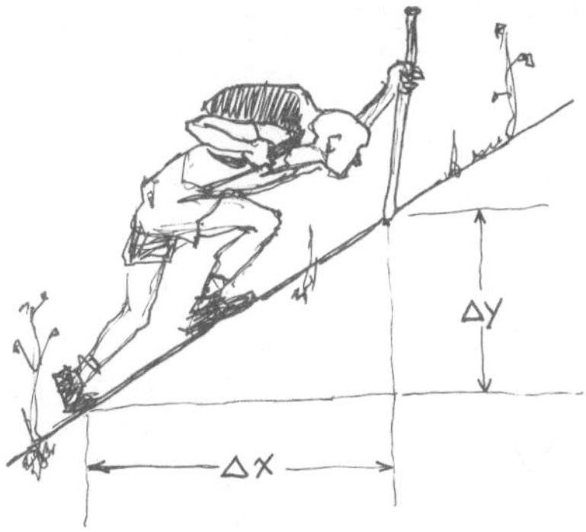

C. *The Derivative*

Definition

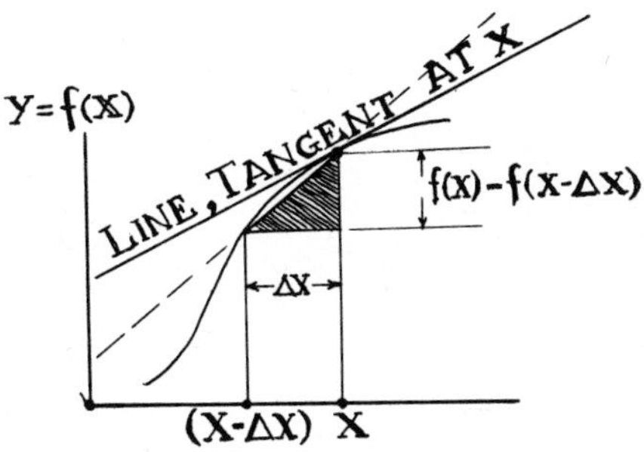

In section A of this chapter it was indicated that the slope as given by the cord approximation approaches the exact slope of the tangent line at a point if Δx_i is allowed to approach zero in the limit.

$$\text{Approximate slope at } x = \frac{\Delta f}{\Delta x} = \frac{f(x) - f(x - \Delta x)}{\Delta x} \qquad \textbf{\textit{(1)}}$$

In accordance with our earlier discussion we *define* the limit,

$$\lim_{\Delta x \to 0} \frac{\Delta f}{\Delta x}, \text{ (if it exists)} \qquad \textbf{\textit{(2)}}$$

to be the slope of the *tangent* line at (x, y); and consequently the tangent line is *defined* to be the line through (x, y) with slope given by ***(2)***. The value of ***(2)*** is usually indicated by

$$\frac{dy}{dx} \text{ or } f'(y) \text{ or } \frac{df}{dx}$$

Caution:

$$\frac{dy}{dx} \text{ is a symbolic abbreviation for } \lim_{\Delta x \to 0} \frac{\Delta y}{\Delta x}.$$

Now $\Delta y/\Delta x$ is a true quotient. However dy/dx is *not* a quotient. That is to say, *we do not define what* dy and dx *are, but only what* $\left(\frac{dy}{dx}\right)$ *is.* Later on we shall give dy, dx independent meanings.

The value of ***(2)*** is called also the *derivative* of f with respect to x, or sometimes the *rate of change* of f with respect to x.

Needless to say, ***(2)*** may very well fail to exist—at least for certain "bad" values of x. It is at any rate clear that if ***(2)*** does exist, then Δy must tend to zero with Δx. That is, we must have

$$\lim_{\Delta x \to 0} \Delta y = 0, \text{ or}$$

$$\lim_{\Delta x \to 0} f(x + \Delta x) = f(x)$$

Consequently, if ***(2)*** *exists for a given* x-value, *then* f(x) *must be continuous at that* x-value. However, it is possible to find examples of functions which are continuous for all values of x but which do not possess a derivative for any value of x. Thus the existence of a derivative is a *stronger* condition than continuity. However, all functions which we shall encounter are "differentiable" except possibly for certain isolated values of x.

Observe: In ***(2)*** we have written $\Delta x \to 0$ and *not* $\Delta x \to O(+)$ or $\Delta x \to O(-)$. That is, Δx may be either > 0 or < 0. The definition ***(2)*** is a two-sided limit. It is possible clearly to define the derivative of f(x) *on the right* (i.e., $\lim_{\Delta x \to O(+)} \frac{\Delta y}{\Delta x}$ and the derivative on the left. We shall not bother about those minor refinements.

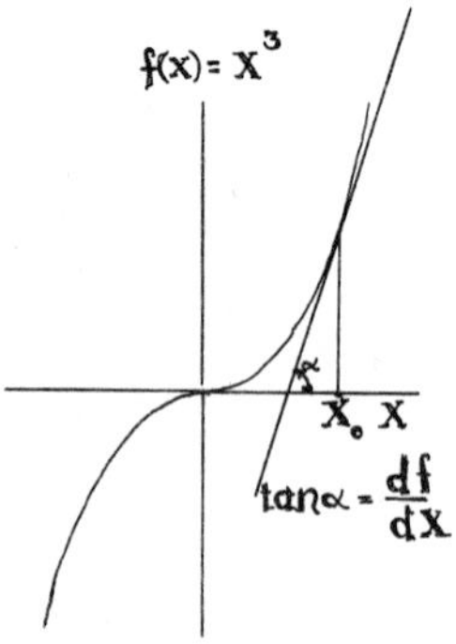

To illustrate the algebraic technique for calculating the derivative of a simple function, consider

$$f(x) = x^3$$

Then

$$\frac{d\,f(x)}{dx} = \lim_{\Delta x \to 0} \left\{ \frac{(x)^3 - (x - \Delta x)^3}{\Delta x} \right\}$$

Expanding this expression we find

$$\frac{df}{dx} = \lim_{\Delta x \to 0} \{3x^2 - 3x\Delta x + (\Delta x)^2\}$$

In the limit the slope clearly tends to a value

$$\frac{df}{dx} = 3x^2$$

We *define* $3x^2$ then to be the slope of $f(x) = x^3$ at the point x.

The tangent line to the curve $f(x) = x^3$ at x_o is the straight line through (x_o, x_o^3) having the slope $3x_o^3$.

In general the function $f = x^n$ where n is a positive integer can be demonstrated to have the derivative

$$\frac{df}{dx} = \frac{d}{dx} x^n = nx^{n-1}$$

The proof of this can be achieved by expanding the ratio

$$\frac{df}{dx} = \lim_{\Delta x \to 0} \left\{ \frac{x^n - (x - \Delta x)^n}{\Delta x} \right\}$$

The expansion of $(x - \Delta x)^n$ is

$$(x - \Delta x)^n = x^n - nx^{n-1} \Delta x + \text{higher terms in } \Delta x$$

Subtracting from x^n and dividing by Δx gives

$$\frac{df}{dx} = \lim_{\Delta x \to 0} nx^{n-1} + \text{terms in } \Delta x)$$

$$= nx^{n-1}$$

Another interesting example is the derivative of sin x and cos x.

$$\frac{d}{dx} \sin x = \lim_{\Delta x \to 0} \left\{ \frac{\sin x - \sin (x - \Delta x)}{\Delta x} \right\}$$

Expand $\sin(x - \Delta x)$

$$\sin(x - \Delta x) = \sin x \cos\Delta x - \cos x \sin\Delta x$$

Remembering that

$$\cos \Delta x \to 1$$
$$\sin \Delta x \to \Delta x$$

For small Δx

then

$$\frac{d}{dx}\sin x = \lim_{\Delta x \to O}\left\{\frac{\sin x - \sin x + \Delta x \cos x}{\Delta x}\right\}$$
$$= \cos x$$

The reader can use the same technique to show that

$$\frac{d}{dx}\cos x = -\sin x$$

Six Golden Rules

It is of great importance to lessen the task of calculating $\frac{d}{dx} f(x)$. This is done most conveniently by deriving formal rules of manipulation for the symbols $\frac{d}{dx}$.

The main formal properties of the operation $\frac{d}{dx}$ are listed below. In 1) – 4) we assume that f(x), g(x) are functions which have derivatives for some value x. That being so, then all the other functions (e.g., f(x) + g(x)) occurring in the formulas also have derivatives at x:

1) $\frac{d}{dx}\{f(x) \pm g(x)\} = \frac{d}{dx}f(x) \pm \frac{d}{dx}g(x)$

2) $\frac{d}{dx}\{c\,f(x)\} = c\frac{d}{dx}f(x)$ (c = constant)

3) $\frac{d}{dx}\{f(x) \cdot g(x)\} = \left(\frac{df}{dx}\right)g + f\left(\frac{dg}{dx}\right)$ (Prod. rule)

4) $\dfrac{d}{dx}\left\{\dfrac{f(x)}{g(x)}\right\} = \dfrac{g(x)\dfrac{df}{dx} - f(x)\dfrac{dg}{dx}}{g^2(x)}$ (quotient rule. If $g(x) \neq O$)

5) $\dfrac{d}{dx} f(u(x)) = \dfrac{df}{du} \cdot \dfrac{du}{dx}$ (chain rule)

6) $\dfrac{dy}{dx} = \dfrac{1}{(dx/dy)}$ (inverse function rule)

Formulas 1) −4) are straightforward rules which tell how to calculate derivatives of sums, products and quotients. Note that 2) is really a special case of 3) if we take g(x) = c. Formulas 4) and 5) will be explained after we show how to prove 1) −4). These really follow very directly from the definition of $\dfrac{df}{dx}$ and from the properties of limits mentioned above. In fact, we leave the demonstration of 1) and 2) as an exercise.

In order to demonstrate rule 3, put

$$f(x - \Delta x) = f(x) - \Delta f$$

and

$$g(x - \Delta g) = g(x) - \Delta g$$

where Δf and Δg represent the change from f(x) and g(x) respectively when the variable changes by Δx.

Then

$$\frac{d}{dx}\{f(x)g(x)\} = \lim_{\Delta x \to O} \left\{\frac{f(x)g(x) - f(x - \Delta x)\, g(x - \Delta x)}{\Delta x}\right\}$$

$$= \lim_{\Delta x \to O} \left\{\frac{f(x)g(x) - [f(x) - \Delta f]\ [g(x) - \Delta g]}{\Delta x}\right\}$$

$$= \lim_{\Delta x \to O} \left\{f\frac{\Delta g}{\Delta x} + \frac{\Delta f}{\Delta x} g - \frac{\Delta f\, \Delta g}{\Delta x}\right\}$$

If Δf and Δg → O as Δx → O then

$$\frac{d}{dx}\{fg\} = f\frac{dg}{dx} + g\frac{df}{dx}$$

Rule 4) is demonstrated in the same manner.

$$\frac{d}{dx}\left\{\frac{f}{g}\right\} = \lim_{\Delta x \to O} \left\{\frac{\left(\frac{f - \Delta f}{g - \Delta g}\right)}{\Delta x}\right\}$$

$$= \lim_{\Delta x \to O} \left\{\frac{1}{g(g - \Delta g)}\left[g\frac{\Delta f}{\Delta x} - f\frac{\Delta g}{\Delta x}\right]\right\}$$

$$= \frac{1}{g}\frac{df}{dg} - \frac{f}{g^2}\frac{dg}{df}$$

Examples

Let n be a whole number (n = 1, 2, 3, etc.). Then for $f(x) = x^{-n} = \frac{1}{x^n}$ we have, applying the quotient rule,

$$\frac{df}{dx}(x) = x^n \cdot \frac{\frac{dl}{dx} - 1 \cdot \frac{d(x^n)}{dx}}{(x^n)^2}$$

Since $\frac{d}{dx}(1) = O$ and $\frac{dx^n}{dx} = nx^{n-1}$, this becomes

$$\frac{d}{dx}(x^{-n}) = -\frac{nx^{-(n-1)}}{x^n \cdot x^n} = -n \cdot x^{-(n-1)}$$

which shows that the rule

$$\frac{d}{dx}x^n = n\,x^{n-1}$$

holds for *all* n $= O, \pm 1, \pm 2, \pm 3$, etc., and for all x if $n \geq O$, for all $x \neq O$ if $n < O$.

The rules so far listed permit us to calculate the derivatives of any polynomial function.

$$f(x) = A_o\,x^n + A_1\,x^{n-1} + \ldots + A_n\ (A_n = \text{constant})$$

Namely, applying 1) repeatedly we have

$$\frac{d}{dx} f(x) = \frac{d}{dx}(A_o x^n) + \frac{d}{dx}(A_1 x^{n-1}) + \ldots + \frac{d}{dx}(A_n)$$

Applying 2), we get

$$\frac{d}{dx} f(x) = A_o \frac{d}{dx}(x^n) + A_1 \frac{d}{dx}(x^{n-1}) + \ldots + A_{n-1} \frac{d}{dx}(x)$$

From the formula above, we obtain finally

$$\frac{d\,f(x)}{dx} = nA_o x^{n-1} + (n-1) A_1 x^{n-2} + \ldots + 2A_{n-2} x + A_{n-1}$$

Thus,

$$\frac{d}{dx}(3x^2 + 4x + 5) = 6x + 4$$

and

$$\frac{d}{dx}(x^3 + x) = 3x^2 + 1$$

As an exercise consider what happens if we apply the operation $\frac{d}{dx}$ to the polynomial $f(x) = A_o x^n + A_1 x^{n-1} + \ldots + A_n$; $(n + 1)$ times in succession.

The quotient rule permits us further to differentiate any *rational* function. E.g.,

$$\frac{d}{dx}\left\{\frac{(3x^2 + 4x + 5)}{x^3 + x}\right\} =$$

$$\frac{(x^3 + x) \cdot \frac{d}{dx}(3x^2 + 4x + 5) - (3x^2 + 4x + 5)\frac{d}{dx}(x^3 + x)}{(x^3 + x)^2}$$

$$= \frac{(x^3 + x)(6x + 4) - (3x^2 + 4x + 5)(3x^2 + 1)}{(x^3 + x)^2}$$

The "Chain Rule" (i.e., rule 5) above) tells how to differentiate a function of a function. Namely, suppose that f(u) is a function of u defined for $a < u < b$, and suppose that $f'(u) = \frac{df}{du}$ exists for some u in that interval. Suppose further that u is made to depend on another variable x, say $u = u(x)$, defined for $\alpha < x < \beta$; and let $\frac{du}{dx}$ exist for an x-value corresponding to the u-value at which $\frac{df}{du}$ is assumed to exist. Then $f = f(u(x))$, regarded as a function of x, has a derivative at that value x, and $\frac{df}{dx} = \frac{df}{du}(u) \cdot \frac{du}{dx}(x)$.

Let $f(u) = y$, and $f(u + \Delta u) = y + \Delta y$

Also put $u(x + \Delta x) = u + \Delta u$

Then $$\frac{f(u(x + \Delta x)) - f(u(x))}{\Delta x} = \frac{f(u + \Delta u) - f(u)}{\Delta x}$$

$$= \left[\frac{f(u + \Delta u) - f(u)}{\Delta u}\right] \cdot \frac{\Delta u}{\Delta x}$$

Let $\Delta x \to O$. Then also $\Delta u \to O$, since Δu is assumed to have a finite limit. Therefore, the above expression tends to $\frac{df}{du} \cdot \frac{du}{dx}$

Further Examples

1. $f(u) = \sin u, \qquad u = 3x^2 + 5$

Then

$$\frac{d}{dx} f(u(x)) = \frac{d \sin u}{du} \cdot \frac{du}{dx}$$

$$= \cos u \cdot \frac{du}{dx}$$

Or

$$\frac{d}{dx} \sin (3x^2 + 5) = [\cos (3x^2 + 5)]\, 6x$$

2. $f(u) = \cos u, \qquad u(x) = \sin x$

Then

$$\frac{df}{dx} = \frac{d}{du} \cos u \cdot \frac{du}{dx} = -\sin u \cdot \cos x$$

Or

$$\frac{d}{dx} \cos(\sin x) = -[\sin(\sin x)] \cos x$$

3. $f(u)$ arbitrary, $u = cx$ (c = constant)

Then

$$\frac{df}{dx} = \frac{df}{du} \cdot \frac{du}{dx} = \frac{df}{du} \cdot c$$

For example

$$\frac{d}{dx} \sin 3x = 3 \cos 3x$$

And

$$\frac{d}{dx} \cos(cx) = -c \sin cx$$

Rule 6) for inverse functions is as follows.

Let $y = f(x)$ be defined and single-valued for $a < x < b$, having a derivative $\neq O$ at some $x = x$ in that interval. Put $y_0 = f(x_0)$. Suppose it is possible to find a *single-valued, continuous* function $g(y)$ for y near y_0 such that $y = f(g(y))$ when y is near y_0. $g(y)$ is none other than one of the possible branches of the "inverse function" obtained by solving $y = f(x)$ for x. Then 6) is to be interpreted as meaning that $x = g(y)$ has a derivative $\frac{dx}{dy}$ at y_0, its value being $1 \Big/ \left(\frac{dy}{dx}\right)$; $\frac{dy}{dx}$ being calculated at x_0. The situation is illustrated below.

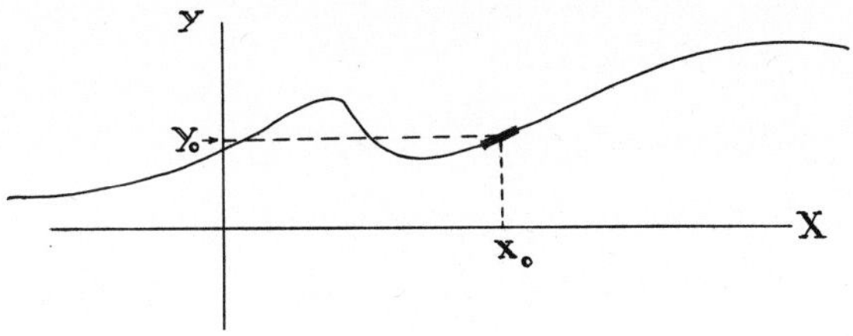

The proof is very simple. We have

$$\frac{g(y+\Delta y)-g(y)}{\Delta y}=\frac{x+\Delta x-x}{\Delta y}=\frac{\Delta x}{\Delta y}=\frac{1}{\frac{(\Delta y)}{\Delta x}}$$

which goes to

$$\frac{1}{\frac{dy}{dx}}$$

Examples:

1.) $y = x^2,$ $\qquad x = \sqrt{y}$

Then

$$\frac{dy}{dx}=2x, \text{ so } \frac{dx}{dy}=\frac{d\sqrt{y}}{dy}=\frac{1}{2x}$$

Or

$$\frac{d\sqrt{y}}{dy}=\frac{1}{2\sqrt{y}}\cdot \text{ i.e., } \frac{d}{dy}(y)^{1/2}=\frac{1}{2}(y)^{-1/2}$$

which is none other than the general rule $\frac{d}{dx}x^n = nx^{n-1}$ with $n = 1/2$ and x replaced by y, which is just as good a letter as x.

2.) $y = \sin x,$ $\qquad x = \text{arc} \sin y$

Then

$$\frac{dx}{dy}=\frac{1}{\frac{dy}{dx}}=\frac{1}{\cos x},\frac{1}{\pm\sqrt{1-\sin^2 x}}$$

Or

$$\frac{d}{dy}\text{arc}\sin y=\frac{1}{\pm\sqrt{1-y^2}}$$

3.) $y = \cos x,$ $\qquad x = \text{arc} \cos y$

$$\frac{dx}{dy}=\frac{1}{-\sin x}=-\frac{1}{\pm\sqrt{1-\cos^2 x}}=\pm\frac{1}{\sqrt{1-y^2}}$$

or

$$\frac{d}{dy}(\text{arc cos } y) = \pm \frac{1}{\sqrt{1 - y^2}}$$

4.) $y = \tan x = \frac{\sin x}{\cos x}$, $\quad x = \text{arc tan } y$

Then

$$\frac{dx}{dy} = \frac{1}{\frac{dy}{dx}}, \text{ and } \frac{dy}{dx} = \frac{\cos^2 x + \sin^2 x}{\cos^2 x} = \frac{1}{\cos^2 x} = \sec^2 x$$

Thus

$$\frac{dx}{dy} = \cos^2 x$$

or

$$\frac{d \text{ arc tan } y}{dy} = \cos^2 x$$

To eliminate x, $y^2 = \frac{\sin^2 x}{\cos^2 x} = \frac{1 - \cos^2 x}{\cos^2 x} = \frac{1}{\cos^2 x} - 1$

or

$$\cos^2 x = \frac{1}{1 + y^2}, \text{ whence the formula}$$

$$\frac{d}{dy}(\text{arc tan } y) = \frac{1}{1 + y^2}$$

Let $y = f(x)$ be defined and single-valued in some interval $a < x < b$. Suppose that f(x) has a *local maximum* at some point x_0 of that interval, meaning simply that $f(x_0 + \Delta x) \leq f(x_0)$ whenever Δx is sufficiently small. Then

$$\frac{f(x_0 + \Delta x) - f(x_0)}{\Delta x} \text{ is } \begin{matrix} \geq O \text{ when } \Delta x < O \\ \leq O \text{ when } \Delta x > O \end{matrix}$$

If $f'(x)$ exists, it follows that its value must be $\geq O$ if $x = (x_0 + \Delta x) \rightarrow x_0$ from the left, but must be $\leq O$ if $x \rightarrow x_0$ from the right. The only possible value for $f'(x_0)$ is therefore zero. Geometrically, this merely says that the tangent to the curve $y = f(x)$ must be horizontal at a local maximum, if the tangent exists to begin with. Similar considerations apply to *local minima.*

Example 1. $f(x) = \sin x$. The maxima or minima can occur only where $f'(x) = \cos x = O$, hence only for

$$x = \pm \pi/2, \pm 3\pi/2, \pm 5\pi/2, \text{etc.}$$

Example 2. $f(x) = (x - 1) \cdot (x^2 + 3x)$

By the product rule, $f'(x) = (x - 1)(2x + 3) + 1 \cdot (x^2 + 3x)$

or $f'(x) = 3x^2 + 4x - 3$

which $= O$ when $x = \dfrac{-4 \pm \sqrt{52}}{6}$

or approximately $x = \dfrac{-4 \pm 7.1}{6} = .52$, and -1.85

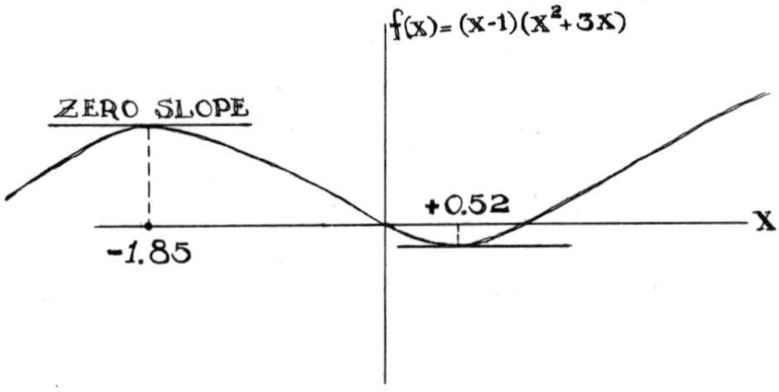

The Mean Value Theorem

Now suppose f(x) is a continuous function in $a \leq x \leq b$ with $f(a) = f(b) = O$, and suppose that $f'(x)$ exists for all x in $a \leq x \leq b$. If f = constant, then f = O, since f = O at the end-points. In this case $f' = O$. If $f \neq$ constant, then f(x) must have either a maximum or a minimum value for some ξ *between* a, b (why?). Since $f'(\xi)$ is assumed to exist, it must be zero. Thus in any case f′(x) must vanish at some point between a, b (perhaps at several, or even all points). This fact is known as *Rolle's theorem.*

Now let f(x) be continuous in an interval $a \leq x \leq b$, and suppose that f′(x) exists at all points x with $a \leq x \leq b$. Put f(a) = A and f(b) = B. Define g(x) by

$$g(x) = B + \frac{(b - x)}{(b - a)}(A - B)$$

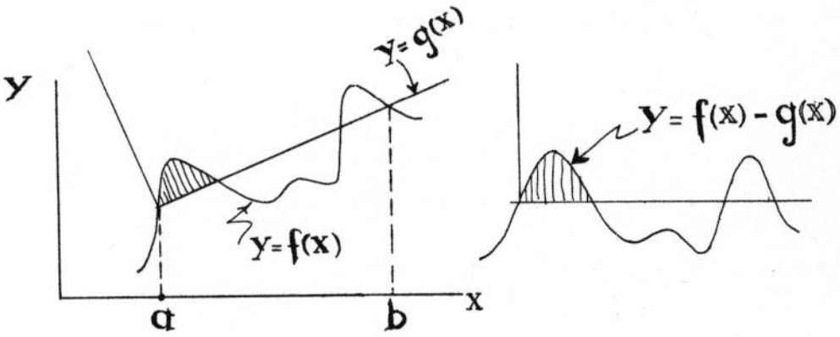

Then clearly f(x) −y(x) satisfies the conditions for Rolle's theorem, and so its derivative $f'(x) - y'(x) = f'(x) + \frac{A - B}{b - a} =$ $f'(x) - \frac{B - A}{b - a}$ must vanish for at least one $x = \xi$ between a, b, i.e.,

$$O = f'(\xi) - \frac{B - A}{b - a}$$

or

$$f(b) = f(a) + f'(\xi) \cdot (b - a) \cdot (a < \xi < b)$$

This is known as the *mean-value* theorem of the differential calculus. From it we conclude at once that

If f(x) *is defined and single-valued in some interval* $x_0 < x < x_1$ *and if* f′(x) *exists and* = O *at all points of that interval (which forces* f(x) *to be continuous),* then f(x) = constant *in that interval.*

For taking $x_0 < a < b < x_1$ we have · , by the M. V. T., f(b) = f(a) + O · (b − a), or f(b) = f(a); and b, being arbitrary, f(x) = f(a) = constant. This result is of great importance.

Higher Derivatives

Once the first derivative has been taken, assuming that it exists, it represents another function. If one takes the derivative (or slope)

of the first derivative the result is called the **second derivative.**

$$\frac{d}{dx}\left(\frac{df}{dx}\right)=\frac{d^2f}{dx^2}$$

In the same manner the derivative of the second derivative is the **third derivative** (or the derivative operation applied 3 times).

$$\frac{d}{dx}\left\{\frac{d}{dx}\left[\frac{df}{dx}\right]\right\}=\frac{d}{dx}\left\{\frac{d^2f}{dx^2}\right\}=\frac{d^3f}{dx^3}$$

The nth derivative is the operation of differentiation applied n **times.**

$$\frac{d}{dx}\left\{\frac{d}{dx}^{\leftarrow n-3 \text{ terms}\rightarrow}\left[\frac{df}{dx}\right]\right\}=\frac{d^nf}{dx^n}$$

As an example, consider

$$f(x) = x^3$$

$$\frac{df}{dx} = 3x^2$$

$$\frac{d^2f}{dx^2} = 6x$$

$$\frac{d^3f}{ax^3} = 6$$

and

$$\frac{d^4f}{dx^4} = O$$

In general

$$\frac{d}{dx}x^n = nx^{n-1}$$

$$\frac{d^2}{dx^2}x^n = n(n-1)x^{n-2}$$

$$\frac{d^m}{dx^m}x^n = n(n-1)(n-2)\text{----}(n-m+1)x^{n-m};\ (m<n)$$

Then

$$\frac{d^m}{dx^m} x^n = O \text{ if } n < m$$

Another example is f(x) = sinx

$$\frac{df}{dx} = \cos x$$

$$\frac{d^2f}{dx^2} = -\sin x$$

and so on.

Maxima and Minima

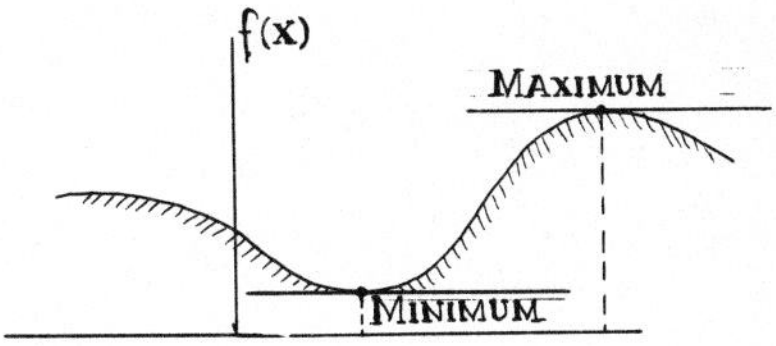

A necessary condition for a maximum or minimum point x_0 in f(x) is that the **slope** of f(x) at x_0 be ZERO.

If the slope vanishes this alone will not provide information as to whether the point of zero slope is a **maximum** or a **minimum point** in the curve f(x).

Maxima have *negative second derivatives.* To see this pictorially plot a maximum point and the region about it.

In addition plot the slope in this interval as a function of x.

Notice that the maximum in f(x) has a corresponding $\frac{df}{dx}$ curve whose slope (i.e., $\frac{d}{dx}\left\{\frac{df}{dx}\right\}$ is negative.

This situation was noted in Chapter 5 when we found that the **curvature matrix** $\mathbb{D}^2$ gave a negative result for curves concave downward.

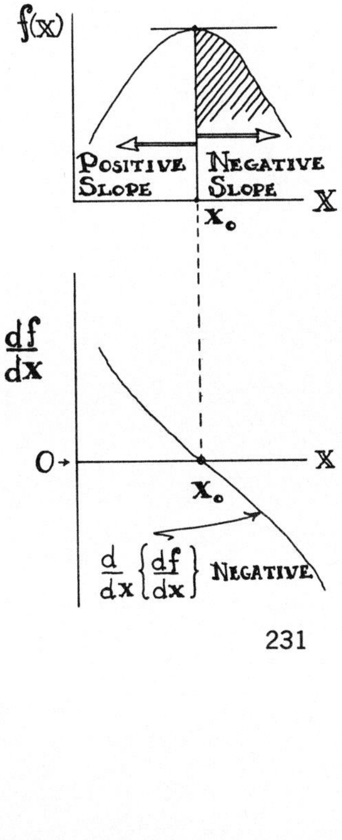

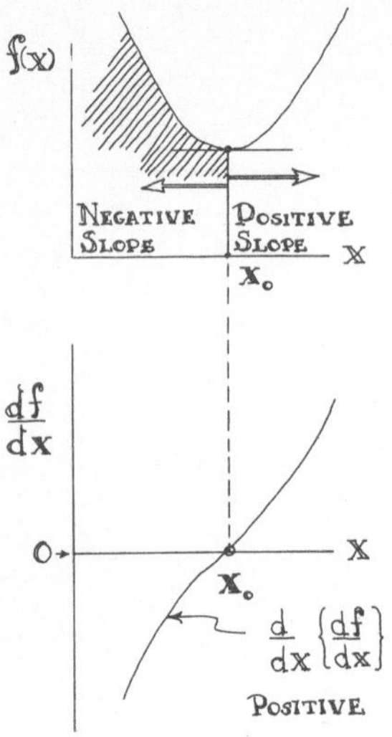

Minima have a *positive* second derivative at the point of zero slope. Again if we plot $\frac{df}{dx}$ against x we find that a minimum point corresponds to a positive slope in $\frac{df}{dx}$ at x_o.

The curvature matrix $\mathbb{D}^2$ gave a positive result for curves concave upward.

Points of inflection. There exists a special case in which the function f(x) has a slope $\frac{df}{dx}$ of **zero** at x_o;

and further

has a zero second derivative. Points such as these are called **points of inflection.**

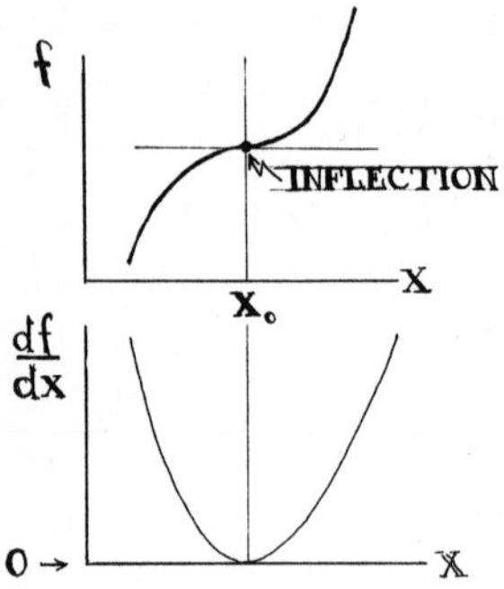

Examples:

Examine $f(x) = \frac{1}{1 + x^2}$

The point at x = O is a point of **zero** slope.

$$\frac{df}{dx} = \frac{-2x}{(1 + x^2)^2}$$

$$\left[\frac{df}{dx}\right] = O$$

at $\qquad x = O$

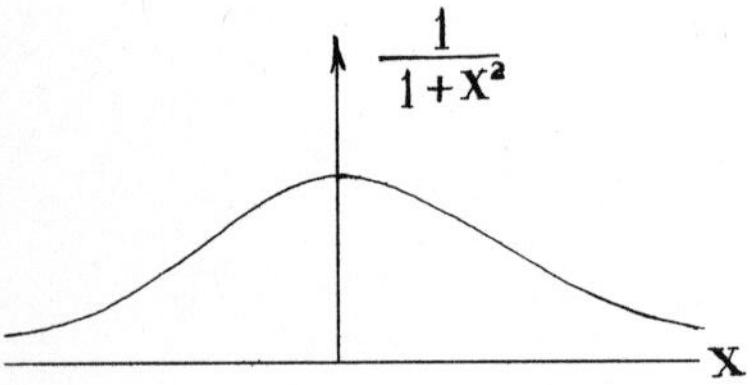

The second derivative is

$$\frac{d}{dx}\left\{\frac{df}{dx}\right\} = \frac{-2}{(1 + x^2)^2} + \frac{2 \cdot 2x(2x)}{(1 + x)^3}$$

$$\left[\frac{d^2f}{dx^2}\right] = -2$$

at $\qquad x = O$

The curve has a maximum at x = O.
Now consider

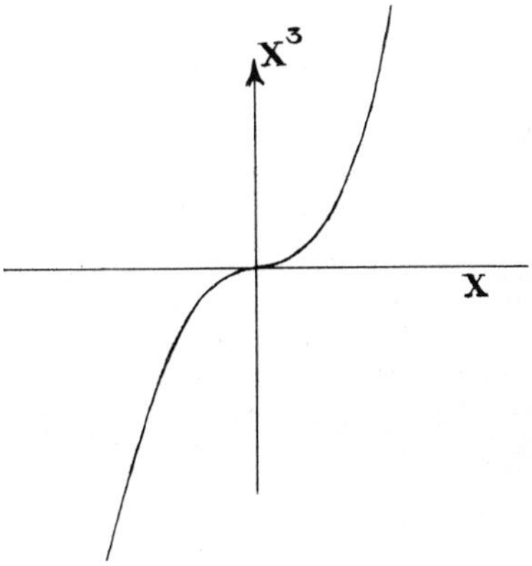

$$f(x) = x^3$$

at $\qquad x = O$

$$\frac{df}{dx} = 3x^2$$

$$\left[\frac{df}{dx}\right] = O$$

at $\qquad x = O$

However

$$\left[\frac{d^2f}{dx^2}\right] = O$$

at $\qquad x = O$

This curve has a point of inflection at the point x = O.

D. The Integral

The Definite Integral

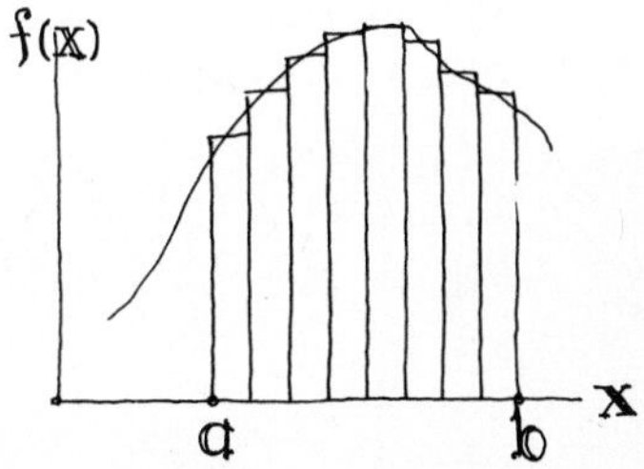

In defining the approximate area under a curve in section A the method of **rectangular partitioning** was utilized.

Our object at this stage is to define the **exact area** under a curve say f(x) in the interval from a to b.

The procedure to follow will be repetitive in that we will review our previous development in somewhat greater detail.

In the diagram to the right we cut the interval (a, b) into say n pieces of length $\Delta x_1, \Delta x_2, \ldots\ldots \Delta x_n$; possibly but not necessarily of equal length.

In each interval we choose a point say $\bar{x}_i$ in the interval Δx_i. $x_{i-1} < \bar{x}_i < x_i$.

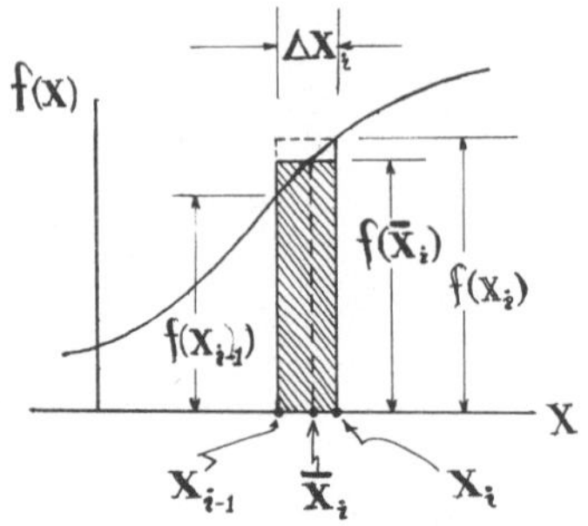

Now construct a rectangle of width Δx_i and height $f(\bar{x}_i)$.

The rectangle provides an approximation to the area under the curve in the ith interval Δx_i, and the area of the rectangle is

$$(\text{height}) \times (\text{width}) = f(\bar{x}_i)\, \Delta x_i$$

We tacitly assume that $f(\bar{x}_i) \geq O$; otherwise of course the area comes out with a **minus** sign. In defining integrals we shall incorporate the possibility of having *negative areas* when $f(\bar{x}_j) < O$.

The total area in the interval a → b in the **rectangle approximation** is then

$$A = \sum_{i=1}^{i=n} f(\bar{x}_i)\, \Delta x_i$$

Again! it is plausible that the value of A approaches some **"Limiting value"** as the subdivisions Δx_i of the interval $a \leq x \leq b$ is made smaller and smaller. As the widths Δx_i are made smaller the number of rectangular sections, n, increases in a manner appropriate to span the interval (a, b)

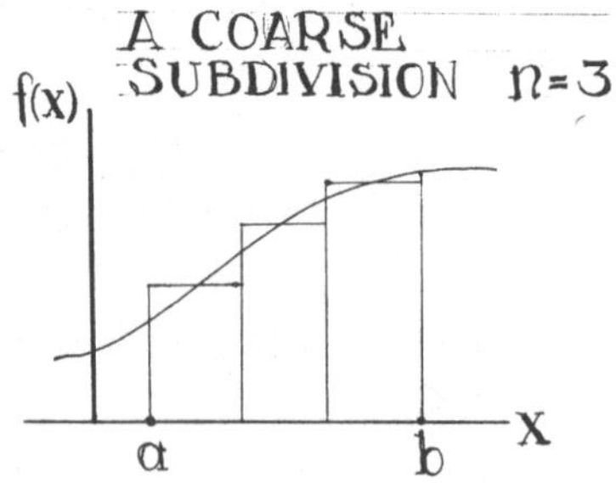

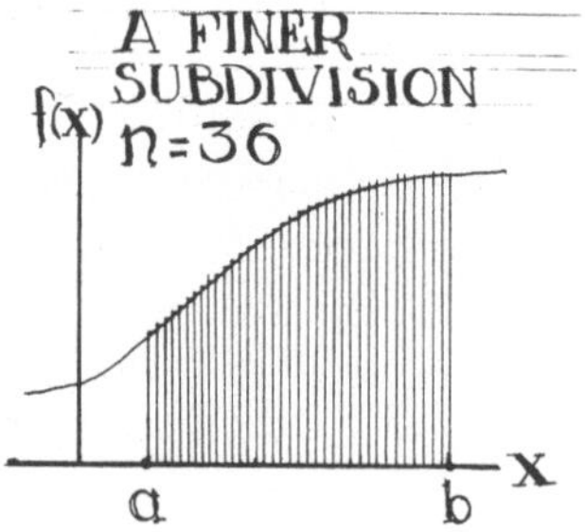

In order to define the **"exact"** area under f(x) in the interval $a \leq x \leq b$ we allow the widths of the subdivisions to approach zero;

$$\Delta x_i \rightarrow O$$

This of course obliges the number of subdivisions n to increase without limit. If N is a number which is arbitrarily large, then

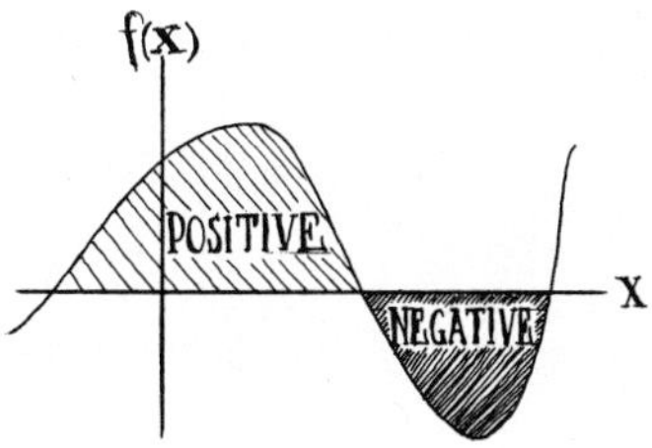

$$\underset{\substack{\Delta x_i \rightarrow O \\ n \rightarrow N}}{\text{Limit}} \sum_{i=1}^{i=n} f(x_i)\,\Delta x_i = \text{The area under } f(x) \text{ in the interval } a \rightarrow b$$

As we mentioned previously *areas* lying under the x axis are to be considered negative.

To exhibit this calculation of the sum in the limit let us utilize the simple curve

$$y = x$$

in the interval $O \leq x \leq b$.

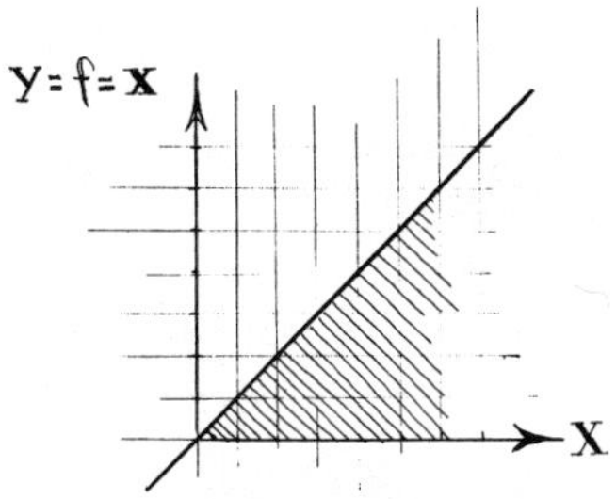

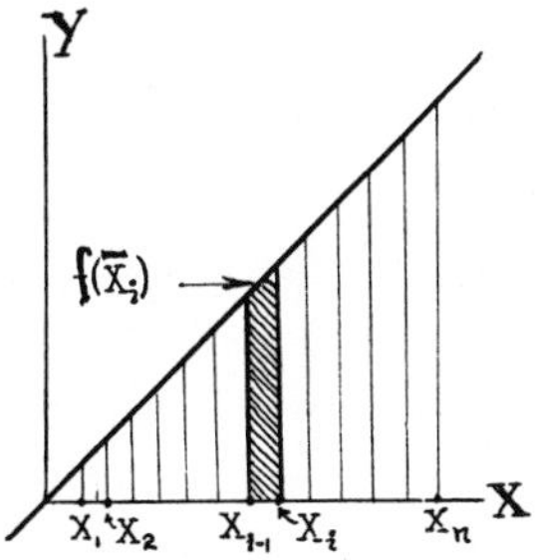

Let the subdivision points be $x_1, x_2 \ldots\ldots x_{n-1}$, putting $x_0 = O$ and $x_n = b$.

Then

$$\Delta x_i = x_i - x_{i-1}$$

and $f(\overline{x}_i) = \overline{x}_i$ for this case. The area from x_0 to x_n is then

$$A \simeq \sum_{i=1}^{i=n} f(\overline{x}_i)\,\Delta x_i = \sum_{i=1}^{i=n} \overline{x}_i\,\Delta x_i$$

$$A = \sum_{i=1}^{i=n} x_i\,(x_i - x_{i-1})$$

Since $x_{i-1} \leq \bar{x}_i \leq x_i$, we have the relation

$$x_{i-1}\Delta x_i \leq \bar{x}_i \Delta x_i \leq x_i \Delta x_i$$

or

$$x_{i-1}(x_i - x_{i-1}) \leq \bar{x}_i (x_i - x_{i-1}) \leq x_i (x_i - x_{i-1})$$

Therefore the area in question lies between

$$A_1 = \sum_{i=1}^{i=n} x_{i-1}(x_i - x_{i-1}), \text{ and } A_2 = \sum_{i=1}^{i=n} x_i (x_i - x_{i-1})$$

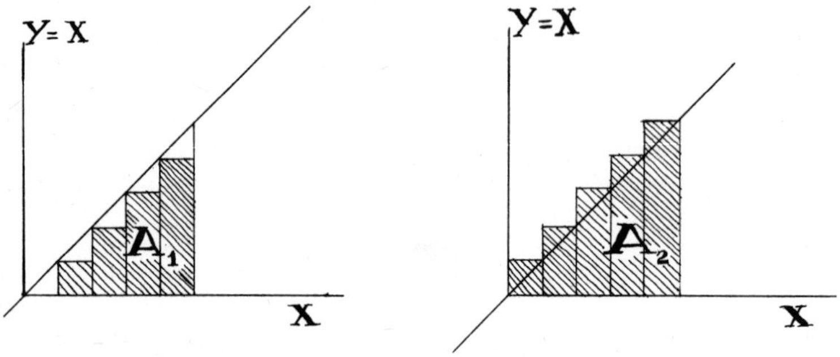

Expanding the terms in the two sums we find

$$A_1 = \sum_{i=1}^{n} x_{i-1} \cdot x_i - \sum_{i=1}^{n} x_{i-1}^2$$

and

$$A_2 = \sum_i x_i^2 - \sum_i x_i \cdot x_{i-1}$$

$$\text{Hence } A_1 + A_2 = \sum_{i=1}^{n} x_i^2 - \sum_{i=1}^{n} x_{i-1}^2$$

$$= (x_1^2 + x_2^2 + \cdots x_n^2) - (x_0^2 + x_1^2 + \cdots x_{n-1}^2)$$

$$= x_n^2 - x_0^2 = b^2 \text{ in this case}$$

The sum $A_1 + A_2$ is independent of the size and number of the subdivisions!

It is interesting to note that the area under the curve $y = x$ in the interval $O \leq x \leq b$ is equal to the **mean** value of the sum of the upper bound A_2 and the lower bound A_1.

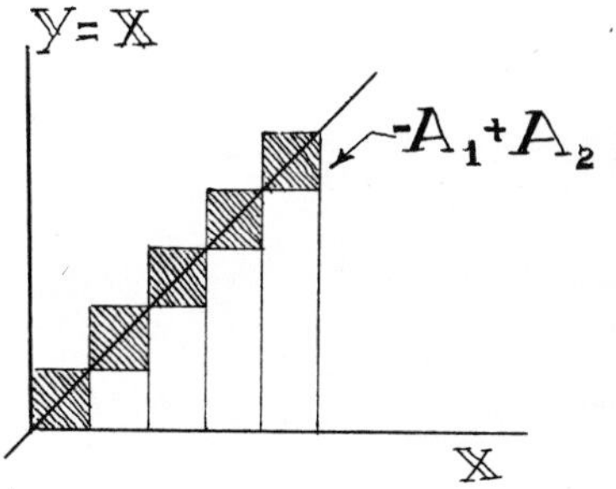

$$\text{Area} = \frac{1}{2}\left\{A_1 + A_2\right\} = \frac{1}{2}b^2$$

The difference $A_2 - A_1$ is also of interest.

$$A^2 - A_1 = \sum_i x_i^2 - 2\sum_i x_{i-1}\,x_i + \sum_i x_{i-1}{}^2$$

$$= \sum_i (x_i - x_{i-1})^2$$

$$= \sum_{i=1}^{n} (\Delta x_i)^2$$

This algebraic result is illustrated graphically above. $(A_2 - A_1)$ is the sum of the shaded squares which is merely the sum of the $(\Delta x_i)^2$.

We wish to see what happens as the Δx_i tend to zero. n of course must at the same time increase without bound. This increase we indicate by the symbol

$$n \to \infty$$

Let h be the greatest of the Δx_i. Then $\Delta x_i \leq h$ for all i, and

$$(\Delta x_i)^2 \leq \Delta x_i\, h$$

whence

$$\sum_{i=1}^{n} (\Delta x_i)^2 \leq \sum_{i=1}^{n} \Delta x_i h = h \sum_{i=1}^{n} \Delta x_i = h\, b$$

This relation provides an inequality for the difference $A_2 - A_1$

$$A_2 - A_1 = \sum_{i=1}^{n} (\Delta x_i)^2 \leq b \text{ x } (\text{max value of } \Delta x_i)$$

Therefore as $\Delta x_i \to O$ (this implies of course that the largest $\Delta x_i \to O$) it follows that

$$A_2 - A_1 \to O$$

Adding $(A_2 + A_1) + (A_2 - A_1) = 2A_2$

we get

$$2A_2 \to b^2 \text{ in the limit, } \Delta x_i \to O$$

Subtracting $(A_2 + A_1) - (A_2 - A_1) = 2A_1$ we get

$$2A_1 \to b^2 \text{ in the limit, } \Delta x_i \to O$$

In conclusion in the limit as $\Delta x_i \to O$

$$\lim_{\Delta x_i \to O} A_1 = \lim_{\Delta x_i \to O} A_2 = A = \frac{1}{2}b^2$$

Since

$$A_1 \leq A \leq A_2, \text{ it follows that}$$

$$A = \frac{1}{2}b^2$$

Accordingly we *define* $A = (1/2)b^2$ as the area under the curve $y = x$ in the interval 0 to b. Clearly this definition gives us the usual area for a triangle.

Needless to say the preceding calculation is cumbersome, and it involves a certain amount of manipulation which is limited to the specific function. One of our main goals at this point is to find a more convenient method for computing areas.

The preceding example however does illustrate in a particularly simple manner the **limiting process.**

Suppose that f(x) is defined and single-valued in $a \leq x \leq b$.

Divide this interval into n parts (not necessarily equal) by points $x_1, x_2, \ldots . x_{n-1}$. Put $a = x_0$ and $b = x_n$, so that we have $x_0\ x_1\ x_2 \ldots\ldots x_n$. The ith subinterval is then $x_{i-1} < x < x_i$. In that interval choose some point $\bar{x}_i$, in the manner that we discussed earlier. The situation is shown below.

The area of the rectangle of height $f(\bar{x}_i)$ and base $\Delta x_i = (x_i - x_{i-1})$ is $f(\bar{x}_i)\ \Delta x_i$. Put

$$S = \sum_{i=1}^{n} f(\bar{x}_i).\ \Delta x_i \qquad (3)$$

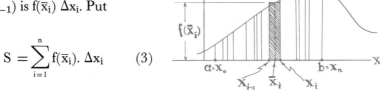

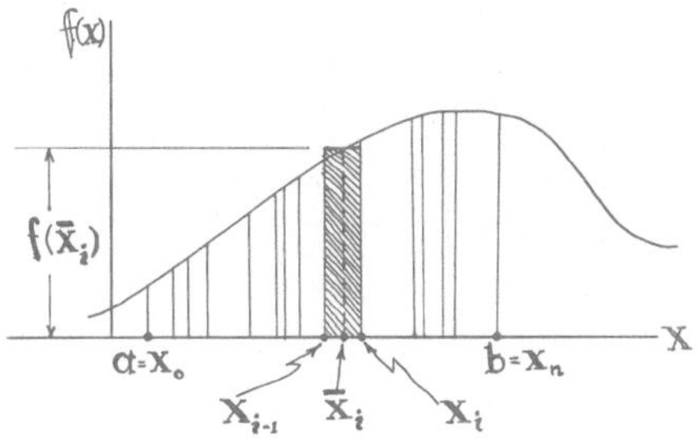

(Such a sum is called a *Riemann sum*). Let $\Delta x =$ max. of $\Delta x_1, \ldots, \Delta x_n$. We consider S as a function of Δx. (It is in general many-valued, because, for one thing, the $\bar{x}_i$ can be chosen in many ways; so can the x_i).

We now define

$$\int_a^b f(x)\,dx = \lim_{\Delta x \to O(+)} S \qquad (4)$$

provided that limit actually exists. The numerical value of (4) we interpret (*by definition*) as the *area* under the curve $y = f(x)$ from $x = a$ to $x = b$, areas under the x-axis being reckoned negative. The dx appearing in (4) is, for the time being, a decoration. The point involved is that x may in turn be considered as a function of a new variable t, say with x running from a to b as t runs from α to β. We can then consider $f(x) = f(x(t))$ as a function of t, and accordingly we can define

$$\int_\alpha^\beta f(x(t))\,dt$$

In general this will not have the same value as (4). Therefore just $\int_a^b f(x)$ would not be an adequate symbol.

A Fundamental Fact: The limit (4) *exists if* $f(x)$ *is continuous in* $a \leq x \leq b$, *and more generally if* $f(x)$ *is continuous save for a finite number of ordinary jump discontinuities in* $a \leq x \leq b$.

We omit the proof of this theorem.

Among the various sums S in (4) are those for which all the Δx_i are equal to $\frac{(b - a)}{n}$, and for which the x_i are chosen to be say at one end-point (x_{i-1} or x_i) or perhaps half way between them. If $\lim_{\Delta x \to O(+)} S$ exists for all possible S, then it must exist for the special ones just mentioned, and therefore in actually calculating integrals it is sufficient to consider only sums S of the special types. For example, our earlier calculation of $\int_O^b x\,dx$ can be done in that way, as follows:

We take

$$x_i = \frac{i \cdot b}{n} (i = O, 1, 2, \ldots, n)$$

so

$$\Delta x_i = \frac{b}{n}$$

We take

$$\bar{x}_i = x_i = \frac{i \cdot b}{n}$$

Then

$$S = f(\bar{x}_i)\,\Delta x_1 + \ldots + f(\bar{x}_n)\,\Delta x_n$$

$$= \frac{1 \cdot b}{n} \cdot \frac{b}{n} + \ldots + \frac{n \cdot b}{n} \cdot \frac{b}{n} = \frac{b^2}{n^2}(1 + 2 + 3 + \ldots n)$$

$$= \frac{b^2}{n^2} n \frac{(n + 1)}{2} = \frac{1}{2} b^2 \frac{(n + 1)}{n} = \frac{1}{2} b^2 \left(1 + \frac{1}{n}\right)$$

Now

$$\Delta x \to O \text{ here means } n \to \infty, \text{ or } \frac{1}{n} \to O \text{ whence}$$

$$S \to 1/2 b^2 \text{ as } \Delta x \to O$$

Consideration of the integral between definite limits a and b defines the

definite integral

which has an interpretation as an area between f(x) and the x axis in the interval a → b.

We shall see in the next section that an indefinite integral can be defined as the inverse operation to that of differentiation.

The Indefinite Integral

That the operation of integration is the inverse to the operation of differentiation has been indicated several times in previous discussions. In particular in section A we demonstrated that the slope of the integral (or area) function at a point was equal to the integrand evaluated at that point.

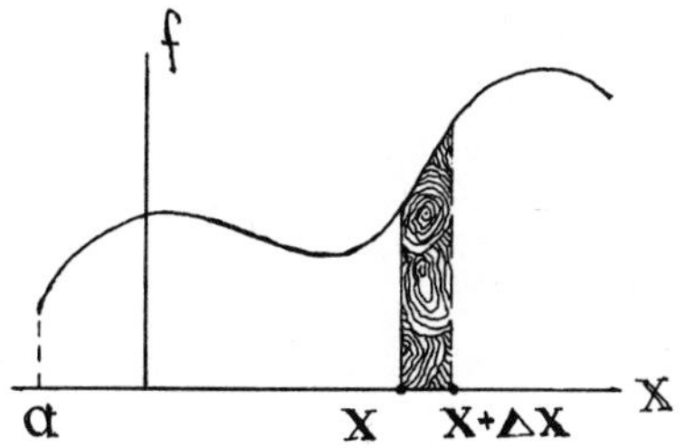

To see this in a slightly different manner consider the definite integral from a to X of f(x)

$$A(X) = \int_a^X f(x)\,dx \qquad (1)$$

Now take the definite integral from a to $(X + \Delta X)$,

$$A(X + \Delta X) = \int_a^{(X+\Delta X)} f(x)\,dx \qquad (2)$$

The area in the small interval ΔX we shall denote as ΔA.

$$\Delta A = A(X + \Delta X) - A(X) = \int_a^{(X+\Delta X)} f(x)\,dx - \int_a^X f(x)\,dx$$

$$= \int_X^{(X+\Delta X)} f(x)\,dx \qquad (3)$$

Note carefully that

$$\int_X^{X+\Delta X} f(x)\,dx$$

is the **area** in the small **trapezoid** having a base ΔX.

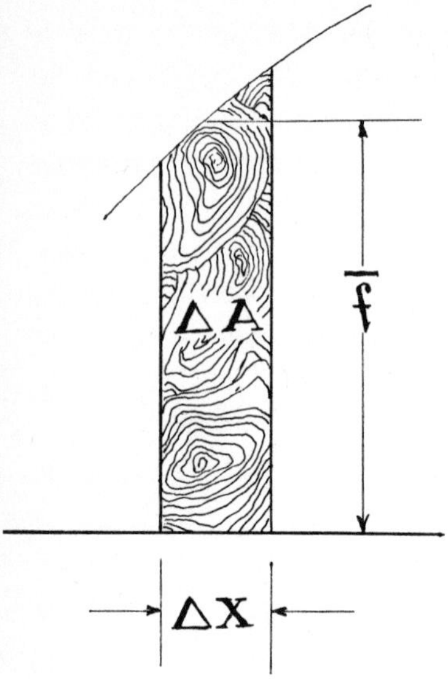

Therefore the average value of f(x) in the interval ΔX is $\bar{f}$.

$$f(X) \leq \bar{f} \leq f(X + \Delta X)$$

where

$$\bar{f} = \frac{\text{area}}{\Delta X} = \frac{1}{\Delta X}\int_X^{(X+\Delta X)} f(x)\,dx \qquad (4)$$

We now divide both sides of equation (3) by ΔX.

$$\frac{\Delta A}{\Delta X} = \frac{1}{\Delta X}\int_X^{(X+\Delta X)} f(x)\,dx = \bar{f}$$

In the limit as ΔX approaches zero $\bar{f}$ approaches f(x); thus

$$\lim_{\Delta X \to O} \frac{\Delta A}{\Delta X} = \frac{dA}{dx} = \lim_{\Delta x \to O} \bar{f} = f(X)$$

Finally if we denote the integral by a function A(x) then the integrand f(x) is equal to the first derivative of the integral.

This relation provides a convenient method for evaluating integrals.

If

$$A(x) = \int f(x')\,dx'$$

then

$$f(x) = \frac{dA\,(x)}{dx}$$

To illustrate the usefulness of this equation let

$$A(x) = x^n$$

then

$$\frac{dA\,(x)}{dx} = nx^{n-1}. = f(x)$$

Because of this, the indefinite integral of nx^{n-1} is

$$\int nx^{n-1}\,dx = x^n \text{ (where } n \neq O)$$

Using this method we see that the definite integrals in most instances can be evaluated from the indefinite integral.

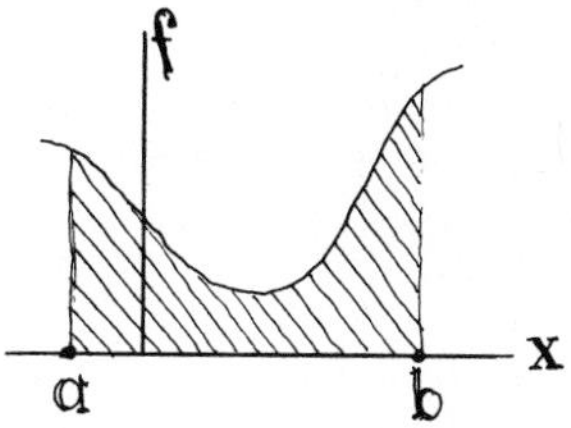

If

$$A(x) = \int f(x')\,dx'$$

then

$$A(b) - A(a) = \int_a^b f(x)\,dx$$

Use our example of $f(x) = nx^{n-1}$ for

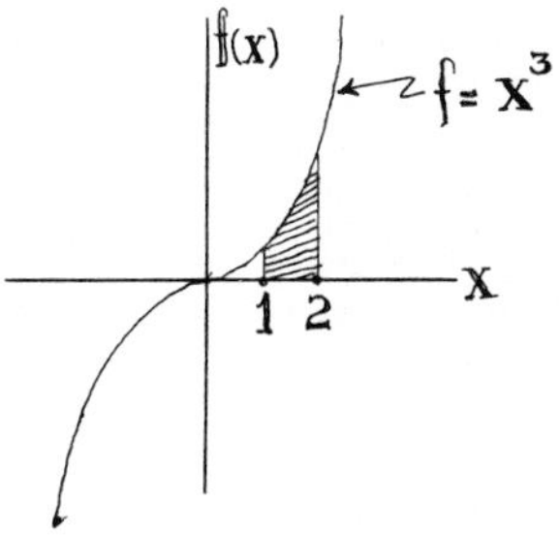

$$n = 4$$

Then

$$A(x) = \int_O^X 4x'^3\,dx' = x^4$$

and considering the area from $x = 1$ to $x = 2$.

$$A(2) - A(1) = (2)^4 - (1)^4 = 16 - 1 = \int_1^2 4x^3\,dx. = 15$$

The area in the interval $x = 1$ to $x = 2$ is then 15.

The case when $n = O$ provides the natural logarithm function which we will take up in a later section.

As another example consider $A(x) = \sin x$;
then

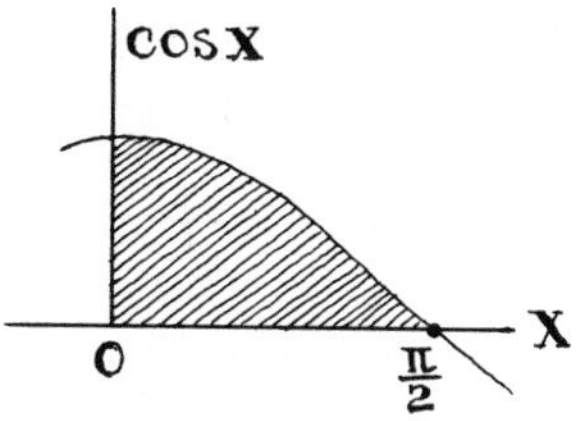

$$\frac{dA}{dx} = \frac{d \sin x}{dx} \cos x$$

and $\qquad A(x) = \sin x = \int \cos x\,dx$

If we now ask for the area under the cosine curve from x = O to x = $\pi/2$

$$A(\pi/2) - A(o) = \sin \pi/2 - \sin O = 1 = \int_O^{\pi/2} \cos x \, dx$$

The reader is encouraged to investigate one or more of the tables of integrals which can be obtained. These tables have large compilations of the integrals for various functions.

The Golden Rules

The so-called **golden rules** for derivatives must have their analogues in the case of integration. In this section the Golden Rules of differentiation will be listed with their inverse relations for integration.

1. $\frac{d}{dx}(F(x) \pm G(x)) = \frac{dF}{dx} + \frac{dG}{dx}$

Then

$$(F(x) + G(x)) = \int \left(\frac{dF}{dx} dx \pm \frac{dG}{dx} dx \right)$$

$$= \int (f(x) \, dx \pm g(x) \, dx)$$

2. $\frac{d}{dx}(c\,F(x)) = c\frac{dF}{dx}$; when c = a constant.

In the same manner

$$\int cf(x) \, dx = c \int f(x) \, dx = c\,F(x)$$

3. The **product rule** gives **integration by parts.**

$$\frac{d}{dx}(F \cdot G) = F \cdot \frac{dG}{dx} + G \cdot \frac{dF}{dx}$$

Then $F(x) \cdot G(x) = \int F\,dG + \int G(x)\,d\,F(x)$
or

$$\int FdG = F(x)\,G(x) - \int G(x)\,dF(x)$$

To illustrate the integration by parts consider the definite integral

$$\int_O^{\pi/2} x \cos x\,dx$$

Let $F(x) = x$ and $dG(x) = \cos x\,dx$ where $G(x) = \sin x$ and $dF =$ dx, then

$$\int_O^{\pi/2} x \cos x = \Big[x \sin x\Big[_O^{\pi/2} - \int_O^{\pi/2} \sin x\,dx = \Big[x \sin x + \cos x\Big[_O^{\pi/2}$$

$$\begin{aligned} &= \pi/2 - O + \cos \pi/2 - \cos O \\ &= \pi/2 - 1 \end{aligned}$$

4. The quotient rule if $G(x) \neq O$

$$\frac{d}{dx}(F/G) = \frac{1}{G}\frac{dF}{dx} - \frac{F}{G^2}\frac{dG}{dx}$$

From this result

$$\int \frac{dF(x)}{G(x)} = \frac{F(x)}{G(x)} + \int \frac{F(x)}{[G(x)]^2}\,dG(x)$$

5. Chain Rule.

$$\frac{d}{dx}\{F(u(x))\} = \frac{dF}{du}\frac{du}{dx}$$

then

$$F(u(x)) = \int \frac{dF(u)}{du}\,\frac{du}{dx}\,dx = \int f(u)\frac{du}{dx}\,dx$$

To demonstrate this rule consider

$$\int \cos (1 + x^2) \cdot 2x \, dx = \sin (1 + x^2)$$

In this example $F(u) = \sin (u)$ and $u = (1 + x^2)$

Examples of integration.

1. $\cos \{\sin x\} = -\int \{\sin (\sin x)\} \cos x \, dx$

2. $\sin 3x = \int 3 \cos 3x \, dx$

3. $\arctan x = \int \frac{dx}{(1 + x^2)}$

Multiple Integrals

Because the function f(x) can be written as an integral our so-called one dimensional integral, which gives the area between f(x) and the x axis in the interval of interest, can be written as a multiple integral.

$$f(x) = \int_0^{f(x)} df'(x)$$

As an example consider

$$f(x) = x$$

then

$$f(x) = \int_0^x dx' = x - 0 = x$$

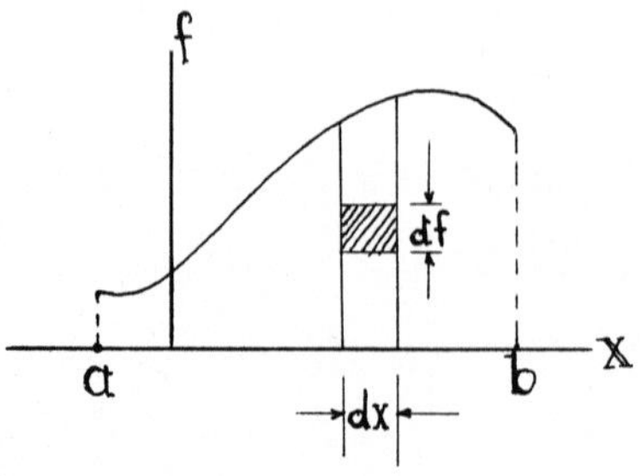

The area between x = a and x = b has been written as

$$A(b) - A(a) = \int_a^b f(x)\,dx$$

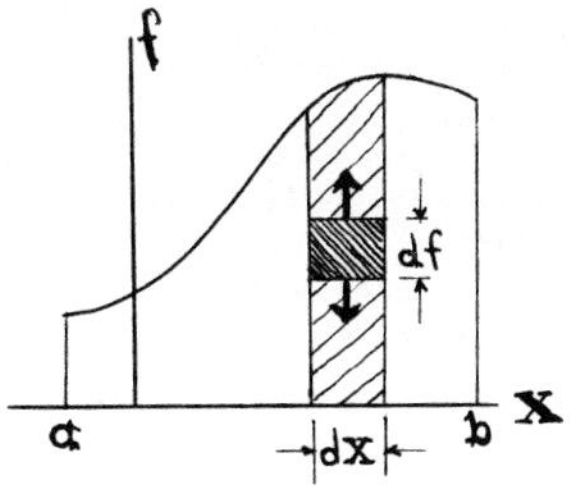

Now we can write this as a multiple integral

$$A(b) - A(a) = \int_a^b \left\{ \int_O^{f(x)} df' \right\} dx = \int_a^b \int_O^{f(x)} df'dx$$

This method of writing the integral implies;

First, that the trapezoidal section in dx be swept out by the operation

$$\int_O^{f(x)} df'$$

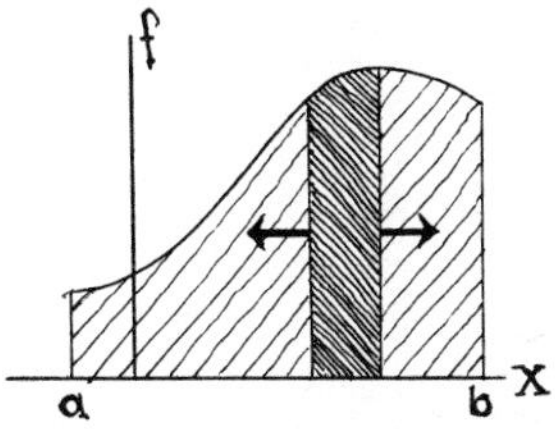

Second, this trapezoid is summed from x = a to x = b.

Note!

We have defined the area between f(x) and the x axis by utilizing a *lower limit* on the df integration of

$$f' = O$$

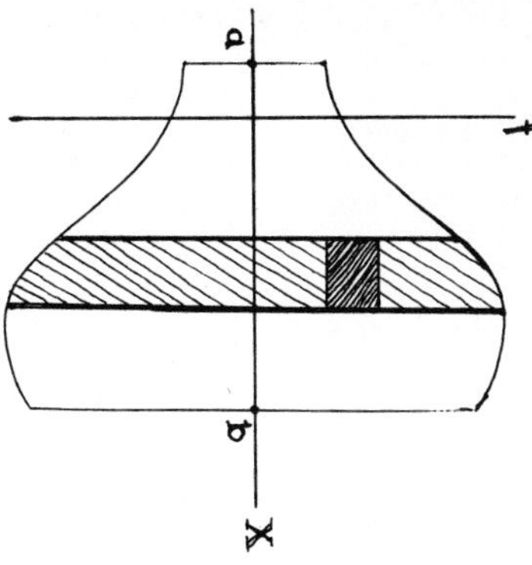

The area between f(x) and the **mirror image** $-f(x)$ could have been obtained by integrating df′ from $-f(x)$ to f(x).

$$\int_a^b \int_{-f(x)}^{f(x)} df'\,dx = \int_a^b 2f(x)\,dx$$

This approach gives twice the area which is to be expected.

We are now in a position to compute areas bounded by certain curves.

Take the curve

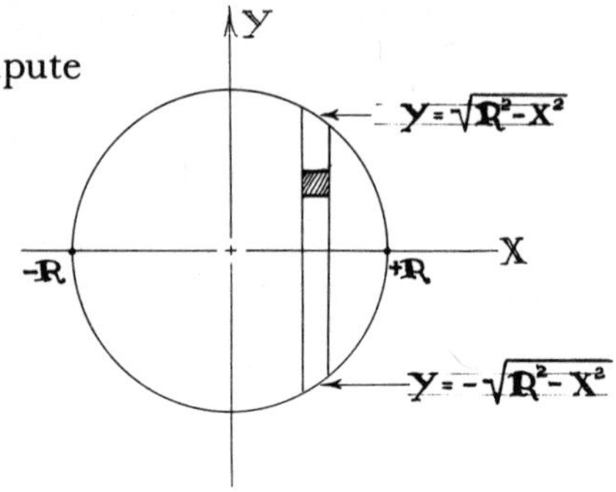

$$x^2 + y^2 + R^2$$

This is a circle, defined by the equation of constraint,

$$y(x) = \pm \sqrt{R^2 - x^2}$$

The area is

$$A_{\text{circle}} = \int_{-R}^{+R}\int_{Y=-\sqrt{R^2-X^2}}^{Y=+\sqrt{R^2-X^2}} dy\,dx = 2\int_{-R}^{+R}\sqrt{R^2 - x^2}\,dx$$

$$= 2R\int_{-R}^{R}(1 - \frac{x^2}{R^2})^{1/2}\,dx$$

Let $x/R = \sin \xi$, then $\frac{dx}{R} = \cos \xi \, d\,\xi$.

When $x = \pm R$; $\sin \xi = \pm 1$; and $\xi = \pm \pi/2$. Substituting into the integral

$$A_{\text{cir}} = 2R^2\int_{-\pi/2}^{\pi/2}\sqrt{1 - \sin^2\xi}\,\cos\xi\,d\xi$$

$$= 2R^2\int_{-\pi/2}^{\pi/2}\cos^2\xi\,d\xi = 2R^2\int_{-\pi/2}^{\pi/2}\frac{1}{2}(1 + \cos 2\xi)\,d\xi$$

$$= \pi R^2$$

In general constraints in two dimensions are provided by equations of the following type,

$$y = y(x)$$

$$\left(\text{for example } y = \pm \sqrt{R^2 - x^2}\right)$$

This problem is equivalent to that of finding the area contained between two curves $y_1(x)$ and $y_2(x)$, and between $x = a$ and $x = b$.

If we ask for the area enclosed between $y_1(x)$ and $y_2(x)$ in the interval a, b the result is

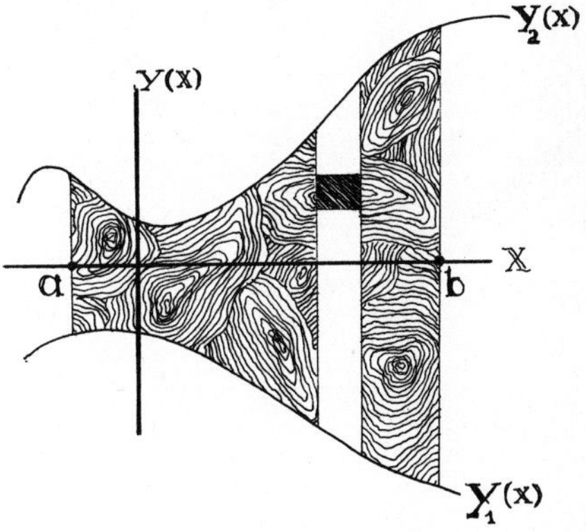

$$\int_a^b \int_{y_1(x)}^{y_2(x)} dx\, dy$$

The order of differential elements indicates that the integration over y should be taken first.

As an example compute the area between

$$y_2 = +x^2$$

and

$$y_1 = -x^2$$

in the interval $x = 0$ to $x = 2$;

then

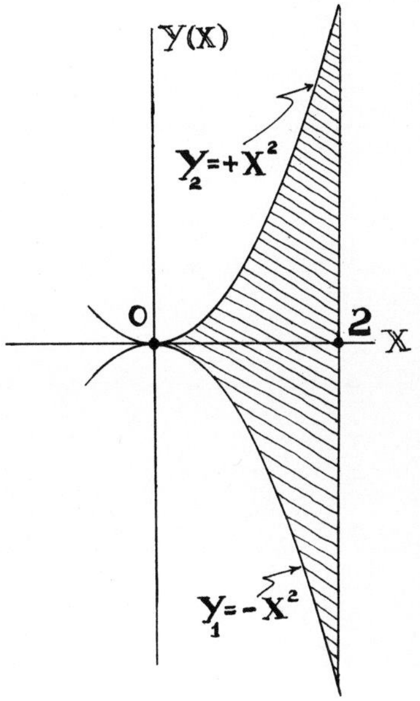

$$\text{Area} = \int_O^2 \int_{-X^2}^{+X^2} dx\, dy =$$

$$\int_O^2 (x^2 - [-x^2])\, dx = \frac{2}{3} x^3 \Big[_O^2 = \frac{16}{3}$$

The computation of volumes is performed in a similar manner.

Assume that we are given a closed surface $z = z(x, y)$ with a boundary condition.

$$z^{min}(xy) \leq z \leq z^{max}(xy)$$

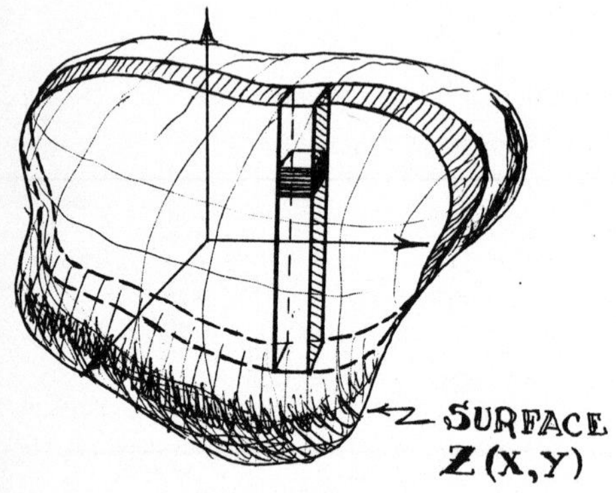

then

$$\text{Volume} = \int_{X_{min}}^{X_{max}} \int_{Y_{min}(X)}^{Y_{max}(X)} \int_{Z_{min}(x,y)}^{Z_{max}(x,y)} dx\, dy\, dz$$

Consider as an example the sphere

$$x^2 + y^2 + z^2 = R^2$$

Then $z = \pm \sqrt{R^2 - x^2 - y^2}$

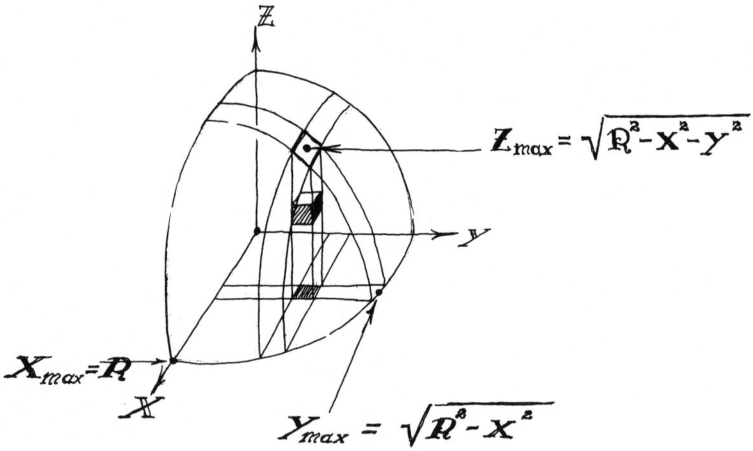

with the projection of the cube dx dy dz on the xy plane having a maximum and minimum value given by

$$y_{min}^{max} = \pm \sqrt{R^2 - x^2}$$

Then

$$\text{Volume} = \int_{-R}^{R} \int_{-\sqrt{R^2-x^2}}^{+\sqrt{R^2-x^2}} \int_{-\sqrt{R^2-x^2-y^2}}^{+\sqrt{R^2-x^2-y^2}} dx\, dy\, dz = \frac{4}{3}\pi R^3$$

In many cases the integrations are simpler in curvilinear coordinates.

For example an area in polar coordinates utilizes an element of area

$$dA = r\, dr\, d\phi$$

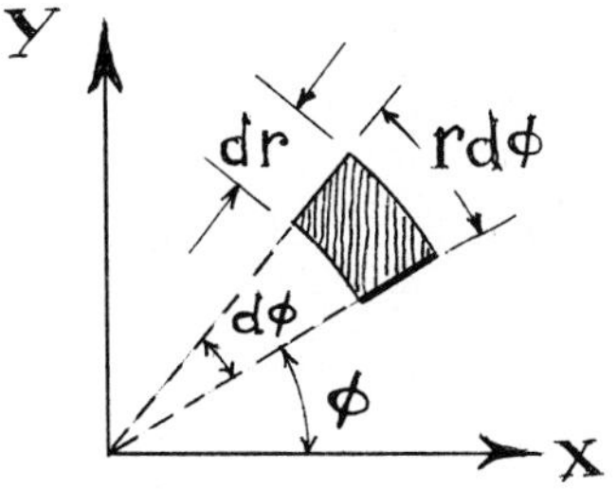

In cylindrical coordinates the volume element is

$$dV = r\, dr\, d\phi\, dz$$

In spherical coordinates the volume element is

$$dV = r^2\, dr \sin\theta\, d\theta\, d\phi$$

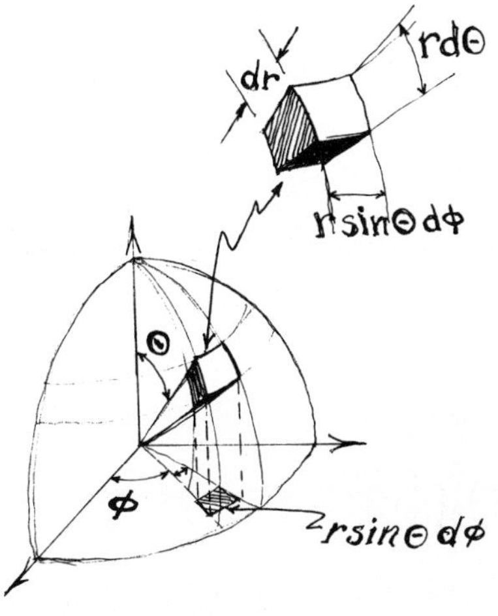

To illustrate the case in which an integral is computed in spherical coordinates let us compute the volume of a sphere.

In spherical coordinates the equation for the spherical surface is

$$r = R = \text{a constant.}$$

Note that the θ and ϕ dependence do not appear.

$$\text{Volume} = \int_0^R \int_0^{\pi} \int_0^{2\pi} r^2\, dr \sin\theta\, d\theta\, d\phi = \frac{4}{3}\pi R^3$$

E. Differential Equations

The solution of differential equations is one of the major everyday tasks of the scientist. The field is large and at most this section of the discussion of the calculus can but give a slight feeling for the subject and some of the basic approaches.

In many cases differential equations are solved by shrewd guesses or insight. In other words one assumes a solution and inserts this solution into the differential equation as a trial solution. Consistency in the final result may be the test of solution. On the other hand some problems are subject to straightforward solution by integration.

An elementary example of this is the problem considered with the matrix operators in Chapter 4. We considered

$$\mathbb{D}\mathbf{f}(x) = \mathbf{g}(x)$$

where

$$\mathbf{g}(x) = x$$

This problem was solved by multiplying by $\mathbb{D}^{-1}$. However, this operation is equivalent to integration

$$\mathbb{D}^{-1} \cdot \mathbb{D} \cdot \mathbf{f} = \mathbf{f} = \mathbb{D}^{-1} \cdot \mathbf{g}$$

In terms of differential operators this equation becomes

$$\frac{d}{dx} f(x) = g(x) = x$$

Note that $\mathbb{D}$ is now $\frac{d}{dx}$

Multiply by dx and integrate (equivalent $\mathbb{D}^{-1}$).

$$\int \frac{df}{dx} dx = \int df = f(x) = \int_{\text{constant}}^{x} x' dx' = \frac{x^2}{2} + \text{constant}$$

Because this section is not intended as a complete exposition it will be limited to the problem of developing the most common functions of mathematics and science, the exponential and trigonometric functions.

When a differential equation is linear the derivatives appear only to the zeroth or first power. Note carefully the difference between *degree* and *power.* The equation

$$\frac{d^2f}{dx^2} + p(x)\frac{df}{dx} + q(x)\,f(x) = O$$

is of **second degree,** however it is **linear.**

The equation

$$\frac{d^2f}{dx^2} + P\left(\frac{df}{dx}\right)^2 + Qf = O, \text{ is of } \textbf{second degree}$$

but is **not linear;** notice that $\frac{df}{dx}$ is taken to the **second power.**

Linear differential equations are subject to solution in series. Whether or not a solution exists at a point depends upon the functions p(x) and q(x) in

$$\frac{d^2f}{dx^2} + p(x)\frac{df}{dx} + q(x)\,f(x) = O$$

This equation has solutions of the type, $f(x) = x^\alpha \sum_{n=O}^{\infty} a_n x^n$.

Linear equations with constant coefficients have exponential or trigonometric solutions.

Consider the differential equation

$$\frac{df}{dx} = f$$

and assume a series solution

$$f = \sum_{n=O}^{\infty} a_n x^n$$

Then

$$\frac{df}{dx} = \sum_{n=1}^{\infty} a_n n x^{n-1} = \sum_{m=O}^{\infty} a_{m+1}(m + 1)\, x^m$$

Substituting into the initial differential equation we obtain

$$\sum_{m=O}^{\infty} \{a_{m+1}(m + 1) - a_m\}\, x^m = O$$

Since the value of x is arbitrary, this equation can only vanish if each coefficient of x vanishes.

Thus

$$a_{m+1} = \frac{a_m}{m + 1}$$

with

$$a_1 = a_0,$$

$$a_2 = \frac{a_1}{2} = \frac{a_0}{1 \cdot 2}$$

$$a_3 = \frac{a_2}{3} = \frac{a_0}{1 \cdot 2 \cdot 3}$$

$$a_n = \frac{a_0}{n!}$$

The solution of this equation is then,

$$f(x) = a_0 \sum_{n=O}^{\infty} \frac{1}{n!} x^n$$

$$= a_0\, e^x$$

The sum defines the exponential to the power x;

$$e^x = \sum_{n=O}^{\infty} \frac{1}{n!} x^n$$

where

$$\frac{de^x}{dx} = \sum_{n=1}^{\infty} \frac{n}{n!} x^{n-1} = \sum_{m=O}^{\infty} \frac{1}{m!} x^m = e^x$$

At this point we should note that this series solution defines the

natural logarithmic integral.

Rearrange the differential equation,

$$\frac{df}{dx} = f$$

to be

$$\int \frac{df}{f} = \int dx = x$$

We know that $f(x) = e^x$ thus

$$x = \log_e e^x = \log_e f(x)$$

Therefore

$$\int \frac{df}{f} = \log_e f(x)$$

Just as the exponential function was derived; the sine and cosine functions are defined by the linear differential equation

$$\frac{d^2g}{dx^2} + g = 0$$

Assume a solution

$$g(x) = \sum_{n=O}^{\infty} b_n x^n$$

then

$$\frac{d^2g}{dx^2} = \sum_{n=2}^{\infty} b_n n(n - 1) x^{n-2} = \sum_{m=O}^{\infty} b_{m+2} (m + 2) (m + 1) x^m$$

Substituting into the original equation we find the relation

$$\sum_{m=O}^{\infty} \{(m + 2) (m + 1) b_{m+2} + b_m\} x^m = O$$

Again the coefficient of x^m must vanish giving,

$$b_{m+2} = \frac{-b_m}{(m + 2) (m + 1)}$$

Two different series are defined from this relation. If we assign a non zero value to b_o all of the even b's are defined;

$$b_2 = \frac{-b_o}{2 \cdot 1}$$

$$b_4 = \frac{-b_2}{4 \cdot 3} = + \frac{b_o}{4 \cdot 3 \cdot 2 \cdot 1}$$

and

$$b_{2n} = \frac{(-1)^n b_o}{(2n)!}$$

This first series is called the **even series**

$$\overset{\text{even}}{g} (x) = b_o \cos x = b_o \sum_{n=O}^{\infty} \frac{(-1)^n}{(2n)!} x^{2n}$$

and defines the cosine (an even function of x);

$$\cos x = \sum_{n=0}^{\infty} \frac{(-1)^n x^{2n}}{(2n)!}$$

An **even** function is defined as a function which *does not* change sign when x is replaced by $-x$.

Note:

$$\cos(-x) = \sum_{n=0}^{\infty} \frac{(-1)^n (-x)^{2n}}{(2n)!} = \sum_{n=0}^{\infty} \frac{(-1)^n x^{2n}}{(2n)!}$$

$$= + \cos x$$

The second series (the odd series) is obtained by setting b_1; then all of the **odd** b's are defined,

$$b_3 = \frac{-b_1}{3 \cdot 2}$$

$$b_5 = \frac{-b_3}{5 \cdot 4} = + \frac{b_1}{5 \cdot 4 \cdot 3 \cdot 2}$$

and

$$b_{2n+1} = \frac{(-1)^n b_1}{(2n+1)!}$$

The **odd series** defines the **sine function** (an odd function of x)

$$g(x)^{\text{odd}} = b_1 \sin x = b_1 \sum_{n=0}^{\infty} \frac{(-1)^n}{(2n+1)!} x^{(2n+1)}$$

An **odd function** of x changes sign when x is replaced by $(-x)$.

$$\sin(x) = \sum_{n=0}^{\infty} \frac{(-1)^n}{(2n+1)!} x^{(2n+1)} = -\sum_{n=0}^{\infty} \frac{(-1)^n (-x)^{(2n+1)}}{(2n+1)!} = -\sin(-x)$$

The reader now has three basic series at his disposal; that for e^x, cos x, and sin x. Using the imaginary $\sqrt{-1} = i$, we are able to prove that

$$e^{ix} = \cos x + i \sin x$$

by *direct substitution* into the series for the exponential. The result should separate into two series; one with the coefficient $+1$, and the other with the coefficient $i = \sqrt{-1}$.

The relation shown above can be also demonstrated except for a sign by showing that e^{ix}

is a solution of
$$\frac{d^2g}{dx^2} + g = 0$$

In closing we should realize in the case of the last equation that there are two **independent** solutions

$$g^{\text{even}} = b_0 \cos x$$

and

$$g^{\text{odd}} = b_1 \sin x$$

The most general solution to this differential equation is a linear combination of the **independent** solutions

$$g(x) = b_0 \cos x + b_1 \sin x$$

The values of b_0 and b_1 are set by the particular problem to be solved.

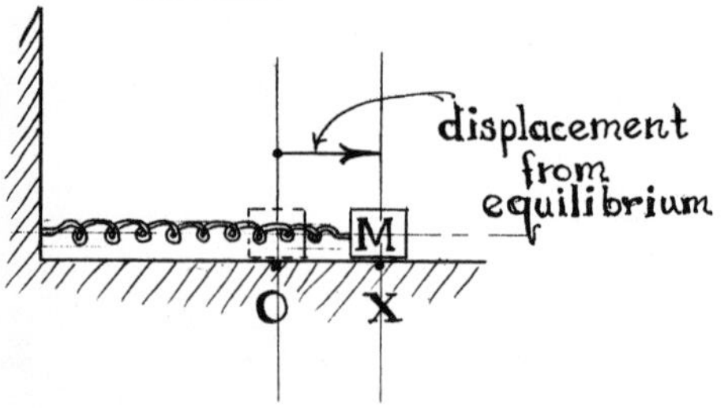

As an example consider a mass M sliding on a frictionless table and connected to a rigid vertical wall by a spring having a spring constant k.

According to Newton's 2nd Law the mass times the acceleration of the body $\frac{(d^2x)}{dt^2}$ is equal to the force F acting upon the body (F = $-kx$, where x = the displacement of the system from equilibrium).

$$M \frac{d^2x}{dt^2} = -kx$$

divide by M and define $\quad \tau = \sqrt{\frac{k}{M}}\, t$

then

$$\frac{d^2x}{d\tau^2} = -x$$

The solution is

$$x(\tau) = b_o \cos \tau + b_1 \sin \tau$$

$$= b_o \cos \left[\left(\frac{k}{M}\right)^{1/2} t\right] + b_1 \sin \left[\left(\frac{k}{M}\right)^{1/2} t\right]$$

If we now stipulate that the system starts from rest ($\frac{dx}{dt} = O$ at $t = O$) with an initial amplitude L, we can determine b_o and b_1. The initial velocity and displacement of the system are called

the initial conditions.

At

$$t = O$$

$$x(O) = L = b_o \cos O + b_i \sin O = b_o$$

$$\frac{dx}{dt}\Big[_{t=O} = O = -b_o \sqrt{\frac{k}{M}} \sin O + b_1 \sqrt{\frac{k}{M}} \cos O$$

$$O = b_1 \sqrt{\frac{k}{M}}$$

Thus

$$b_o = L$$

and

$$b_1 = O$$

Our solution with these initial conditions is

$$x(t) = L \cos \sqrt{\frac{k}{M}}\, t$$

One further remark should be included. The reader will notice the similarity between this problem and say the **eigenvalue** problem in two dimensions (the ellipse). Both problems have a set of **independent** solutions; two eigenvectors in the case of the ellipse and two independent functions in the case of the differential equation.

The two orthogonal **eigenvectors** of the ellipse problem form a set of **base vectors** in two dimensions. As a result any arbitrary vector in that space can be represented in terms of its projection along the two bases.

The two independent solutions of the differential equation in a sense form a set of bases in the respect that any arbitrary configuration of the system described (of the spring system in our particular example) can be expanded in terms of the two base functions.

Think about it.

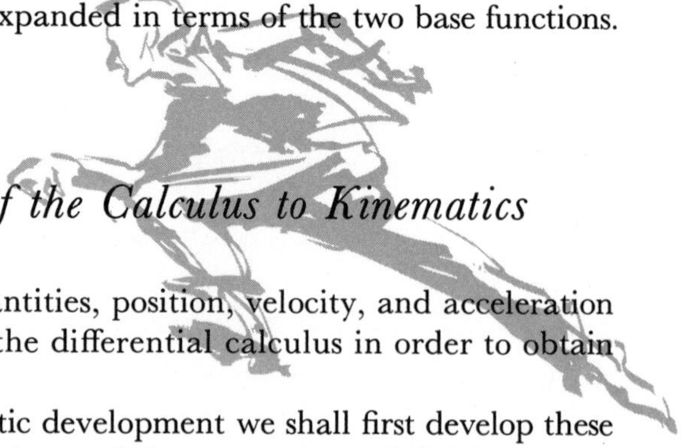

F. Applications of the Calculus to Kinematics

The kinematical quantities, position, velocity, and acceleration require explicit use of the differential calculus in order to obtain proper definition.

To provide a systematic development we shall first develop these quantities in one dimension and then extend our definition to two dimensions. The formulation of velocities and accelerations in three dimensions is a simple extension of the two dimensional problem.

To further simplify the problem we shall use as an *example* to illustrate the technique the motion of a particle with *constant acceleration.*

Velocity in One Dimension

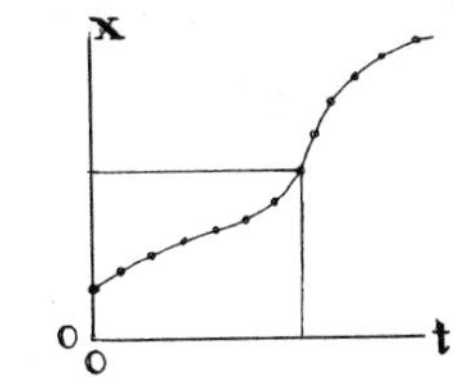

Suppose that we are observing an object moving in a straight line, and that we obtain the distance of the object from an arbitrary origin O as a function of time.

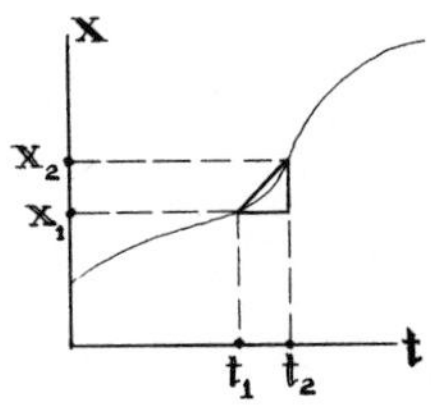

The average velocity in a time interval $(t_2 - t_1)$ is defined as

$$v_{ave} = \frac{x(t_2) - x(t_1)}{(t_2 - t_1)}$$

The path of the object in space-time is shown in the graph above. We see that the average velocity is the chord approximation for the slope of the x(t) curve in the interval $(t_2 - t_1)$.

The **instantaneous velocity** at a given time t_1 can be obtained in the limit as the interval $(t_2 - t_1)$ goes to zero.

Let $(t_2 - t_1) = \Delta t$

Then $t_2 = t_1 + \Delta t$

and

$$v = \lim_{\Delta t \to O} \frac{x(t_1 + \Delta t) - x(t_1)}{(t_1 + \Delta t) - t_1} = \lim_{\Delta t \to O} \frac{x(t_1 + \Delta t) - x(t_1)}{\Delta t}$$

or

$$v = \frac{dx}{dt}$$

The instantaneous velocity is the slope of the space-time curve. To illustrate this definition first consider a body moving with constant velocity

$$x(t) = x_0 + v_0 t$$

Then

$$v = \frac{dx}{dt} = \frac{d}{dt}(x_0 = v_0 t) = v_0$$

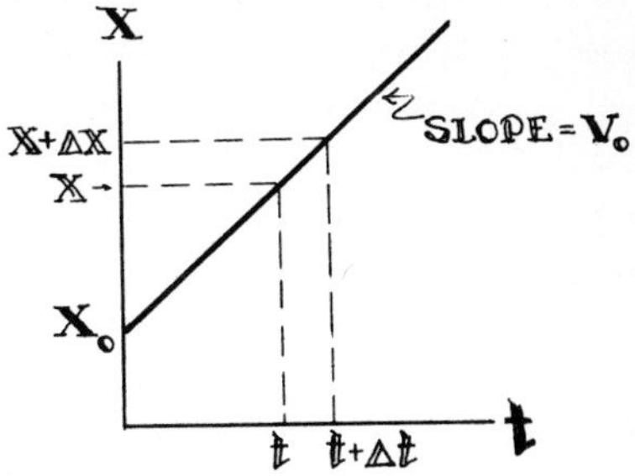

Another simple example arises when we consider the curve x(t) which is quadratic in the time

$$x(t) = x_0 + v_0 t + \frac{1}{2} a_0 t^2$$

In this case

$$v(t) = \frac{dx}{dt} = v_0 + a_0 t$$

In general we can plot a corresponding velocity-time curve, and in this simple case the velocity is linear in time.

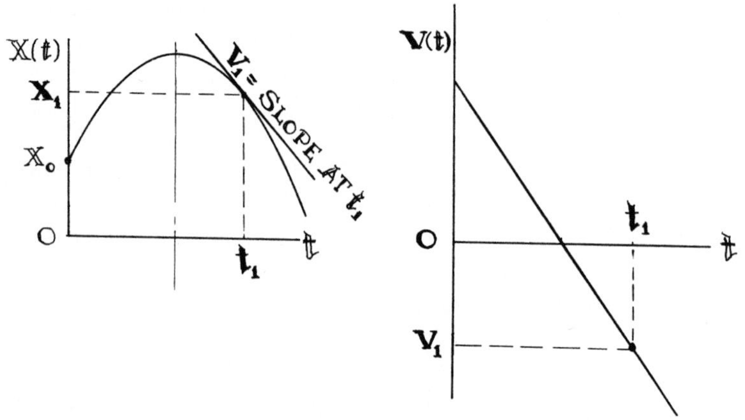

Acceleration in One Dimension

The acceleration is defined as the time rate of change of the velocity. In order to compute the acceleration we must use the velocity-time curve as shown below. Obviously if only the x(t) curve is given we must obtain the v(t) by differentiation.

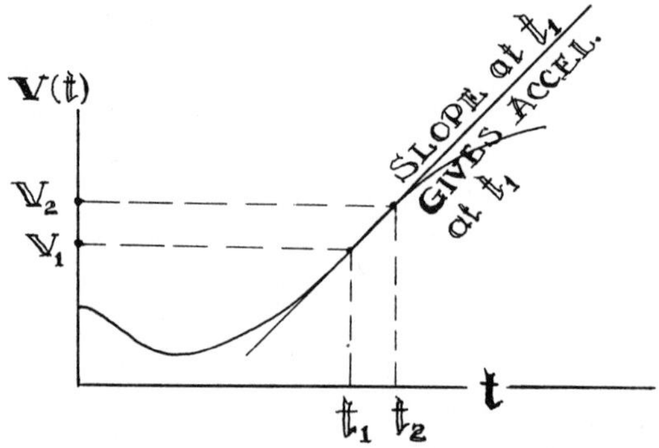

The **average acceleration** a in the time interval $(t_2 - t_1)$ is defined as

$$a_{ave} = \frac{v(t_2) - v(t_1)}{(t_2 - t_1)}$$

As in the case of the velocity the **instantaneous acceleration** "a" at a time say t_1 is obtained by letting the time interval $\Delta t = (t_2 - t_1)$, approach zero in the limit

$$a = \lim_{\Delta t \to O} \frac{v(t + \Delta t) - v(t)}{(t + \Delta t) - t} = \lim_{\Delta t \to O} \frac{v(t + \Delta t) - v(t)}{\Delta t}$$

or

$$a = \frac{dv}{dt}$$

The acceleration is the slope of the velocity curve in one dimension.

Because $v = \frac{dx}{dt}$ we can write

$$a = \frac{d}{dt}\{v\} = \frac{d}{dt}\{\frac{dx}{dt}\} = \frac{d^2x}{dt^2}$$

and we see that the acceleration is the second derivative of the space-time curve x(t).

Again we can use the previous examples to illustrate the calculation of the acceleration.

In the case of the function

$$x(t) = x_o + v_o t \text{ and}$$

$$v(t) = \frac{dx}{dt} = v_o = \text{a const}$$

$$a(t) = \frac{dv}{dt} = O$$

This is the function representing motion with *constant velocity* and therefore *zero acceleration.*

The second example,

$$x(t) = x_0 + v_0 t + \frac{1}{2} a_0 t^2$$

and $$v(t) = v_0 + a_0 t$$

giving $$a(t) = \frac{dv}{dt} = a_0 = \text{a constant}$$

Thus this function represents motion with constant acceleration.

Displacement in Two Dimensions

The graphical representation of the motion of a particle in two dimensions is not as easily displayed as in one dimension. If $\mathbf{r}_2$ and $\mathbf{r}_1$ represent the position of a particle at times t_2 and t_1 respectively, then the displacement in the time interval $(t_2 - t_1)$ is

$$\mathbf{r}_2 - \mathbf{r}_1 = \mathbf{r}(t_2) - \mathbf{r}(t_1)$$

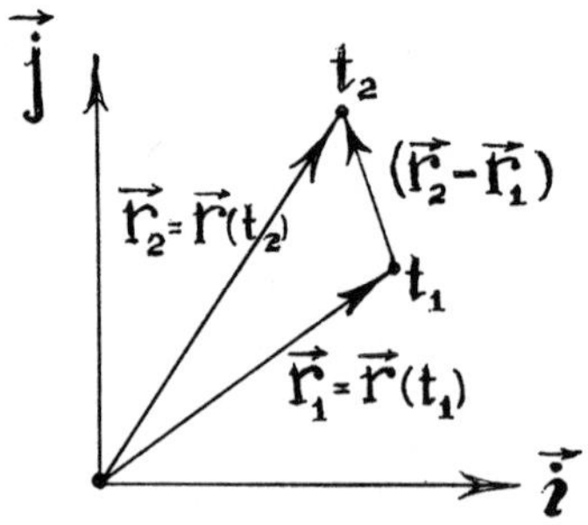

Since

$$\mathbf{r} = \mathbf{r}(t) = x(t)\,\mathbf{i} + y(t)\,\mathbf{j}$$

We can write

$$\mathbf{r}_2 - \mathbf{r}_1 = \{x(t_2) - x(t_1)\}\,\mathbf{i} + \{y(t_2) - y(t_1)\}\,\mathbf{j}$$

Velocity in Two Dimensions

Strictly speaking the term **velocity** is a vector quantity having both direction and magnitude. The magnitude of the velocity is called **the speed.**

Thus in the previous one dimensional problem we were actually computing the speed.

The diagram in section (a) above can be utilized to compute the average velocity in the time interval $(t_2 - t_1)$ in a manner similar to that used for the one dimensional problem.

$$\mathbf{v}_{ave} = \frac{\mathbf{r}_2 - \mathbf{r}_1}{t_2 - t_1} = \frac{\mathbf{r}(t_2) - \mathbf{r}(t_1)}{t_2 - t_1}$$

$$= \left\{ \frac{x(t_2) - x(t_1)}{t_2 - t_1} \right\} \mathbf{i} + \left\{ \frac{y(t_2) - y(t_1)}{t_2 - t_1} \right\} \mathbf{j}$$

$$= v_x \mathbf{i} + v_y \mathbf{j}$$

The **instantaneous vector velocity** is defined by letting the time interval $(t_2 - t_1) = \Delta t$ and then to let $\Delta t \to O$ in the limit.

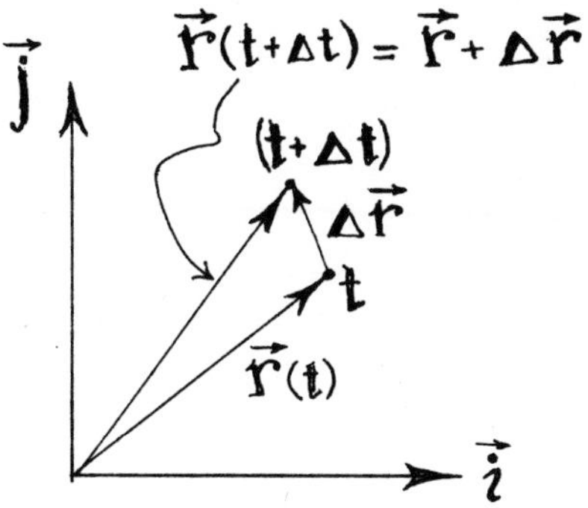

$$\mathbf{v} = \lim_{\Delta t \to O} \frac{\mathbf{r}(t + \Delta t) - \mathbf{r}(t)}{(t + \Delta t) - t} =$$

$$\lim_{\Delta t \to O} \frac{\mathbf{r}(t + \Delta t) - \mathbf{r}(t)}{\Delta t}$$

Thus

$$\mathbf{v} = \lim_{\Delta t \to O} \frac{\Delta \mathbf{r}}{\Delta t} = \frac{d\mathbf{r}}{dt} = \left(\frac{dx}{dt}\right)\mathbf{i} + \left(\frac{dy}{dt}\right)\mathbf{j}$$

The final form in terms of the time derivatives of the components x and y assumes that the base vectors **i** and **j** are held constant in time.

It is possible to define the components of **r** in terms of a set of rotating base vectors. Under such conditions we must also consider the time derivatives of the bases.

The time dependent position vector representing motion with constant velocity $\mathbf{v}_0$ is

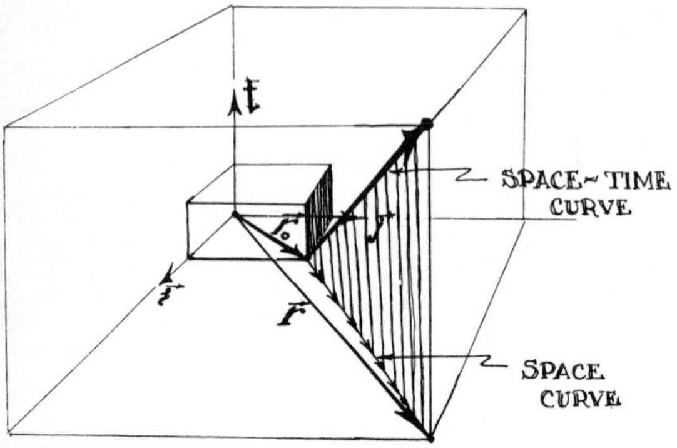

$$\mathbf{r}(t) = \mathbf{r}_0 + \mathbf{v}_0 t$$

Then

$$\frac{d\mathbf{r}}{dt} = \mathbf{v}_0$$

$$\mathbf{v}_0 = v_{ox}\,\mathbf{i} + v_{oy}\,\mathbf{j} = \text{a constant vector}$$

Motion with constant acceleration is represented by

$$\mathbf{r}(t) = \mathbf{r}_0 + \mathbf{v}_0 t + \frac{1}{2}\mathbf{a}_0 t^2$$

here $\mathbf{r}_0$, $\mathbf{v}_0$, and $\mathbf{a}_0$

are all constant vectors.

Let

$$\mathbf{r}_0 = O; \text{ (starts from the origin)}$$
$$\mathbf{v}_0 = (v_0 \cos \alpha)\mathbf{i} + (v_0 \sin \alpha)\mathbf{j}$$

and

$$\mathbf{a}_0 = -g\mathbf{j}$$

The equation for **r** then represents the position of a projectile fired with an initial velocity $\mathbf{v}_0$ and in a direction inclined at an angle α relative to the horizontal.

The acceleration will be shown to be equal in magnitude to g the acceleration of gravity, and the acceleration is directed in the negative **j** direction.

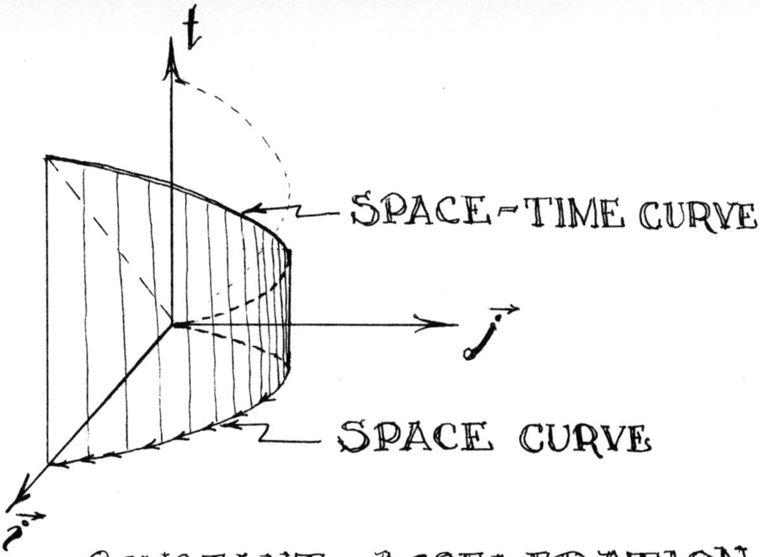

CONSTANT ACCELERATION

Solving for the velocity in this special case we find

$$\frac{d\mathbf{r}}{dt} = \mathbf{v}_0 + \mathbf{a}_0 t$$

$$= \{v_0 \cos \alpha\}\,\mathbf{i} + \{(v_0 \sin \alpha) - gt\}\,\mathbf{j}$$

We note that the particle moves in the **i** direction with a constant velocity ($v_0 \cos \alpha$).

The velocity in the y direction is linear in the time.

The use of base vectors enables us to keep track of the two components of the motion with one set of symbols.

Acceleration in Two Dimensions

The acceleration is defined in the velocity-time space. Again we recognize the advantages of our unit vectors **i** and **j**. Because they are dimensionless and serve to indicate direction only, the velocity space can be super-posed upon the position space.

The average acceleration in a time interval $(t_2 - t_1)$ is

$$\mathbf{a}_{ave} = \frac{\mathbf{v}(t_2) - \mathbf{v}(t_1)}{(t_2 - t_1)}$$

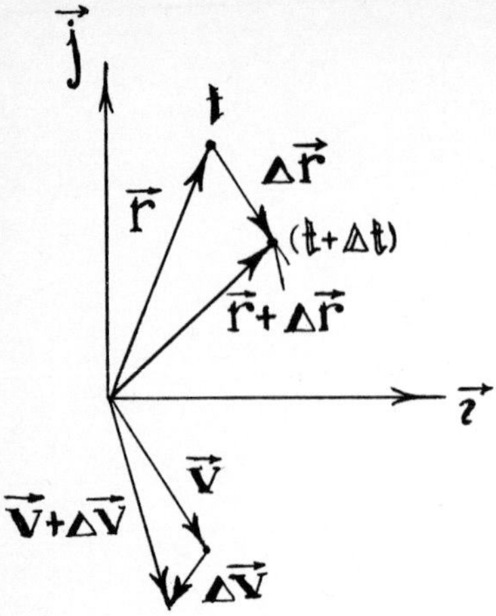

As we let the time interval become arbitrarily small we define the **instantaneous acceleration** at a time t

$$\mathbf{a} = \lim_{\Delta t \to O} \frac{\mathbf{v}(t + \Delta t) - \mathbf{v}(t)}{(t \times \Delta t) - t}$$

$$\mathbf{a} = \lim_{\Delta t \to O} \frac{\Delta \mathbf{v}}{\Delta t} = \frac{d\mathbf{v}}{dt} = \left(\frac{d^2x}{dt^2}\right)\mathbf{i} + \left(\frac{d^2y}{dt^2}\right)\mathbf{j}$$

Since
$$\mathbf{v} = \frac{d\mathbf{r}}{dt}$$

$$\mathbf{a} = \frac{d}{dt}\left\{\frac{d\mathbf{r}}{dt}\right\} = \frac{d^2\mathbf{r}}{dt^2} = \left\{\frac{d^2\mathbf{x}}{dt^2}\right\} i + \left\{\frac{d^2y}{dt^2}\right\} \mathbf{j}$$

In our example of motion with constant velocity $\mathbf{v}_o$.

$$\mathbf{r}(t) = \mathbf{r}_o + \mathbf{v}_o t$$

$$\mathbf{v}(t) = \frac{d\mathbf{r}}{dt} = \mathbf{v}_o = \text{a constant vector}$$

and

$$\mathbf{a}(t) = O$$

Our second example dealt with a projectile fired from the origin of coordinates ($\mathbf{r}_o = O$) with an initial velocity $\mathbf{v}_o = (v_o \cos \alpha)\mathbf{i} + (v_o \sin \alpha)\,\mathbf{j}$.

$$\mathbf{r}(t) = \mathbf{v}_o\, t + \frac{1}{2}\mathbf{a}_o\, t^2$$

$$\mathbf{v} = \frac{d\mathbf{r}}{dt} = \mathbf{v}_o + \mathbf{a}_o t$$

where

$$\mathbf{a}_o = -g\,\mathbf{j} \qquad [g = 32.2 \text{ ft/sec}^2, \text{ or } g = 980 \text{ cm/sec}^2]$$

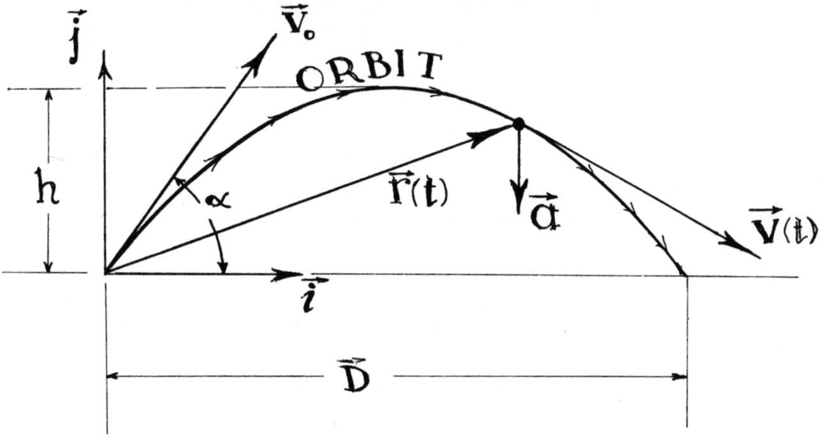

The instantaneous acceleration is

$$\mathbf{a}\,(t) = \frac{d\mathbf{v}}{dt} = \mathbf{a}_o = -g\,\mathbf{j}$$

and is a constant vector, directed downward.

We note that $\mathbf{v}$ (t) is tangent to the orbit $y = y(x)$.

The **orbit** equation can be obtained by eliminating the parameter t in x(t) and y(t).

Since

$$\mathbf{r}\,(t) = x(t)\,\mathbf{i} + y\,(t)\,\mathbf{j} = (v_o \cos\alpha)\,\mathbf{i} + \{v_o \sin\alpha - \frac{1}{2}gt^2\}\,\mathbf{j}$$

Equating coefficients of $\mathbf{i}$ and of $\mathbf{j}$ we obtain

$$x(t) = (v_o \cos\alpha)\,t$$

$$y(t) = (v_o \sin\alpha)\,t - \frac{1}{2}gt^2$$

By eliminating the parameter t in these two simultaneous algebraic equations we obtain

$$y = (\tan\alpha)\,x - \frac{1}{2}g\left(\frac{x}{v_o \cos\alpha}\right)^2$$

The maximum height h to which the projectile rises is obtained by finding the time t_h at which the y component of v(t) vanishes

$$v_y = v_o \sin\alpha - gt$$

$$O = v_o \sin\alpha - gt_h$$

giving

$$t_h = \frac{v_o \sin\alpha}{g}$$

Then

$$h = y\,(t_h) = \frac{1}{2}\frac{(v_o \sin\alpha)^2}{g}$$

The maximum distance D traveled in the x direction is of course obtained by finding the time t_o at which y is *zero*.

Since

$$y\,(t) = (v_o \sin\alpha)\,t - \frac{1}{2}gt^2$$

$$O = \{v_o \sin\alpha - \frac{1}{2}gt_D\}\,t_D$$

One solution is t = O since the particle starts at y = O. The other solution is

$$t_D = \frac{2v_o \sin\alpha}{g}$$

Then

$$x(t_D) = D = 2v_o{}^2\,\frac{\sin\alpha\cos\alpha}{g}$$

Index

E

F

G

H

I

K

L

M